Energy Models
in Emerging Economies

About the Centre

The Centre for Science and Technology of the Non-Aligned and Other Developing Countries (NAM S&T Centre) is an inter-governmental organisation with a membership of 47 countries spread over Asia, Africa, Middle East and Latin America. Besides this, 11 S and T agencies and academic/research institutions of Bolivia, Brazil, India, Nigeria and Turkey are the members of the S and T-Industry Network of the Centre. The Centre was set up in 1989 to promote South-South cooperation through mutually beneficial partnerships among scientists and technologists and scientific organisations in developing countries. It implements a variety of programmes including international workshops, meetings, roundtables, training courses and collaborative projects and brings out scientific publications, including a quarterly Newsletter. It is also implementing 7 Fellowship schemes, namely, Research Training Fellowship for Developing Country Scientists (RTF-DCS), Joint NAM S&T Centre – ZMT Bremen Fellowship, Joint NAM S&T Centre – ICCBS Fellowship, Joint CSIR/CFTRI (Diamond Jubilee) – NAM S&T Centre Fellowship, Joint NAM S&T Centre – DST (South Africa) Training Fellowship on Minerals Processing and Beneficiation, NAM S&T Centre Research Fellowship, NAM S&T Centre – U2ACN2 Research Associateship in Nanosciences and Nanotechnology in Indian institutions. These activities provide, among others, the opportunity for scientist-to-scientist contact and interaction, training and expert assistance, familiarising the scientific community on the latest developments and techniques in the subject areas, and identification of technologies for transfer between member countries. The Centre has so far brought out 79 publications and has organised 108 international workshops and training programmes.

For further details, please visit www.namstct.org or write to the Director General, NAM S&T Centre, Core 6A, 2nd Floor, India Habitat Centre, Lodhi Road, New Delhi-110003, India (Phone: +91-11-24645134/24644974; Fax: +91-11-24644973; E-mail: namstcentre@gmail.com; namstct@bol.net.in).

Energy Models in Emerging Economies

— Editor —

Sasi K Kottayil

CENTRE FOR SCIENCE & TECHNOLOGY OF THE
NON-ALIGNED AND OTHER DEVELOPING COUNTRIES
(NAM S&T CENTRE)

2018
DAYA PUBLISHING HOUSE®
A Division of
ASTRAL INTERNATIONAL PVT. LTD.
New Delhi – 110 002

© 2018 EDITOR
ISBN: 9789388173650 (Int. Edn.)

Published by   : **Daya Publishing House®**
                         A Division of
                         Astral International Pvt. Ltd.
                         – ISO 9001:2015 Certified Company –
                         4736/23, Ansari Road, Darya Ganj
                         New Delhi-110 002
                         Ph. 011-43549197, 23278134
                         E-mail: info@astralint.com
                         Website: www.astralint.com

Digitally Printed at : **Replika Press Pvt. Ltd.**

Preface

The richest 20 per cent of the world population consumes about sixty per cent of the commercial energy produced globally. While the richest ten per cent of the world population is responsible for about fifty per cent of global lifestyle consumption emissions,the poorest 50 per cent has a meager share of only 10 per cent (Ecologize.in, 9 December 2015);yet the latter will be the worst affected by the imminent climate change. Thus it is the predicament of the developing or underdeveloped nations today to shoulder the dual responsibilities of fighting the impended natural calamities on one side and fast implementing the much needed fuel switching on to sustainable energy resources. The fifteen articles presented in this book narrate the effects of climate change in ten evolving economies of Asia and Africa as well as the efforts taken by their governments to plan future energy. Part I includes nine articles on *Energy Planning*, while Part II has six on *Energy Research and Development*.

The first article in Part I establishes with proof that Nigerian climate is changing and the reasons behind the change are perspicuously identified. Methodologies for evolving new energy models in developing countries have been suggested in the next, also from Nigeria, on the realisation that 90 per cent of net energy demand growth until 2035 is expected to come from emerging economies (while it is only 57 per cent now). The third article identifies opportunities, challenges and strategies for emission reduction following systematic and historical studies, from Indonesia. The fourth one details how Cuba with current fossil fuel share in energy above 95 per cent targets for 2030 renewable energy portfolio to the tune of 24 per cent. Technology transfer is identified as a viable energy development model and demonstrates priority analysis for it in the fifth article, from Turkey. The next one describes the Renewable Energy Master Plan of Nigeria. Nigeria's Federal Ministry of Environment aims to increase the contribution of Renewable Energy to account for 10 per cent of Nigerian total energy consumption by 2025.

Vietnam's National Green Growth Strategy is presented in the seventh article. It examines resource mobilization strategy for the required green growth. A critical analysis of National Urban Transport Policy of India can be read in the next article. Green building rating mechanism in India and its recent innovations are discussed in the last article of Part I.

Part II presents research perspectives and a few R&D activities in the energy sector. Influence of weather on electricity consumption in two West African cities has been observed and analysed in the first article from Togo. Carbon foot print of Indian telecom sector is exposed and green alternatives are suggested in the next article. The third article from Iran envisages sustainable biohydrogen economy and presents a detailed account of relevant technologies. An Indian perspective of ICT based resilient management solutions for urban transport infrastructure has been presented in the fourth article. The fifth one is from Sri Lanka and it tells about an innovative biogas digester. The last is an academic study from Egypt on off-grid generation of electricity from wind.

That the authors are either researchers or policy makers in their countries claims more than authenticity the intimacy to ground realities. Beyond energy statistics and technology these articles serve readers, expectedly policy makers and researchers, with glimpses of life and governance in emerging economies. These articles have originally been presented as papers in the fourth international conference, on the theme same as the title of this book, jointly organised by NAM S&T Centre, India and SEEM, India at Ahmedabad, India during 12-14 December 2016. Delegates from sixteen countries, all energy and environment professionals, attended the presentations and interacted with the authors in the conference. Reviewed, revised and edited, the selected fifteen papers formed the articles contained by this book. Clinging to the mission of spreading the message of clean, green and lean energy to the global brains, the NAM S&T Centre – SEEM team dedicates this to you, earnestly.

Sasi K. Kottayil

Introduction

Climate change is a big challenge to the mankind. It is well known that the earth's atmosphere is growing warmer due to Greenhouse Gas (GHG) emissions generated by human activity and extensive use of fossil fuels in every sphere of life which is leading to visible climate change across the globe and is also threatening to wipe out the human presence in many regions of the world. The 21st Conference of the Parties – COP 21 - to the United Nations Framework Convention on Climate Change (UNFCCC) held in Paris in 2015 reaffirmed the target of keeping the rise in temperature below 2°C. The most obvious remedial measures to counter the global warming comprise enacting and following a comprehensive strategy for accelerating the transition from fossil fuel consumption to increasing the share of renewable energy in the overall energy use; researching innovative efficient ways to utilise the energy in everyday life and sectors like manufacturing, transport, agriculture *etc.* and simultaneously, exploring ways to adapt to the impacts of climate change. Thus there is a need to evolve and use the energy models, policy frameworks and guidelines by the governments and concerned stakeholders. The development of the policy framework for Energy Models is intended to help provide the rapidly evolving process of policy making with a much needed roadmap. Ultimately, the purpose of the framework is to support processes that protect, and enhance, human well-being in the face of climate change.

Evolving Energy Models is about practice rather than theory, commencing with the information already possessed by the countries in climatically vulnerable systems such as agriculture, water resources, public health and disaster management, and aims at exploiting the existing synergies and intersecting themes. These Energy Models can be used by the countries to both evaluate and complement the existing planning processes to address climate change adaptations. Wherever needed, these may be either freshly formulated or may be prepared to complement the existing models, guidelines and policies.

In order to exhaustively deliberate on above issues, the Centre for Science and Technology of the Non-Aligned and Other Developing Countries (NAM S&T Centre) organised the 4th **Triennial International Workshop** on "**Energy Models in Emerging Economies - COP 21**" in Ahmedabad, Gujarat, India during 12-14 December 2016 in partnership with the Society of Energy Engineers and Managers (SEEM), India.

The Workshop was attended by 23 senior experts and professionals from 13 NAM countries, including Cuba, Egypt, India, Indonesia, Iran, Malaysia, Nigeria, Sri Lanka, Togo, Turkey, Vietnam, Zambia and Zimbabwe, and the USA.

As a follow up of the Ahmedabad workshop, the papers submitted by the participants were compiled and have been brought out in the form of the present book – '**Energy Models in Emerging Economies**– which has been edited by Dr. Sasi K Kottayil, President, Society of Energy Engineers and Managers (SEEM), INDIA. The book contains 15 scientific/technical papers contributed by the experts from 10 countries.

The publication of the book has been possible due to the commitment and valuable efforts of the entire team of the NAM S&T Centre, especially of Dr. Kavita Mehra, Mr. M. Bandyopadhyay and Ms. Meenu Galyan at all the stages of the publication process. The contribution of Mr. Pankaj Buttan in designing of the cover page, formatting and liaising with the printers is worthy of mention.

I am sure that this book will be useful to the policy experts and technology service providers, business, civil societies and multi-lateral agencies amongst developing countries on ensuring universal access to modern energy services, financing their clean energy sources as well as taking measures of adaptation to the effects of global warming within the perspective of 'Emerging Energy Models' initiative.

Prof. Dr. Arun P. Kulshreshtha
Director General,
NAM S&T Centre

Contents

Part II: R&D in Energy

Part I

Energy Planning

Chapter 1

Effect of Climate Change in Nigeria

Sule Yakubu Okolo

Chief Scientific Officer,
Raw Materials Research and Development Council,
Plot 17, Aguiyi Ironsi Street, Maitama,
PMB 232, Garki, Abuja, Nigeria

Abstract

Climate change refers to change in average weather conditions on Earth over a long period of time. The scientific community agrees that the change is due mostly to human use of fuels and other activities that release carbon dioxide and other greenhouse gases into the atmosphere. Nigeria is one of the world's most densely populated countries with a population of 180 million people, half of which are considered to be in abject poverty. Nigeria is recognized as being vulnerable to climate change. Climate change and global warming if left unchecked will cause adverse effects on livelihoods in Nigeria, such as crop production, livestock production, fisheries, forestry and post-harvest activities, because the rainfall regimes and patterns will be altered, floods which devastate farmlands would occur, increase in temperature and humidity which increases pest and disease would occur and other natural disasters like floods, ocean and storm surges, which not only damage Nigerians' livelihood but also cause harm to life and property, would occur.

Keywords: NIMET, Carbon sink, Environment protection, Gas flaring.

Introduction

The **Federal Republic of Nigeria** commonly referred to as **Nigeria**, is in West Africa, bordering Benin in the west, Chad and Cameroon in the east, and Niger in the north. Its coast in the south lies on the Gulf of Guinea in the Atlantic Ocean. It comprises 36 states and the Federal Capital Territory, where the capital, Abuja is located. Nigeria lies between longitudes 2°49' E – 14°37E and latitudes 4°16'N-13°

52'N and is in the humid tropics. It has a land area of 923, 850 km². Nigeria is often referred to as the "Giant of Africa", owing to its large population and economy. With approximately 184 million inhabitants, Nigeria is the most populous country in Africa.

Climate change refers to change in average weather conditions on Earth over a long period of time. The scientific community agrees that the change is due mostly to human use of fuels and other activities that release carbon dioxide and other greenhouse gases into the atmosphere. Other causes of climate change include such factors like changes in solar energy received by the earth, biotic processes, tectonic and volcanic activities within the earth crust. Earth temperature depends on the equilibrium between the energy entering and leaving the planet earth. When the inward energy from the sun is absorbed by the earth, it gets warmed and when the absorbed energy is reflected back into the space the earth gets cooled.

Global climate change has already had observable effects on our environment. Glaciers have shrunk, ice on rivers and lakes is breaking up earlier, plant and animal ranges have shifted and trees are flowering sooner. Effects that scientists had predicted in the past would result from global climate change are now occurring: loss of sea ice, accelerated sea level rise and longer, more intense heat waves due to greenhouse gases produced by human activities. Nigeria, like every other African country, is experiencing unfavourable climatic conditions, with negative impacts on the welfare of its citizens.

The effects of activities of oil companies in the Niger Delta region of Nigeria are felt in the rural agricultural areas, which compound the climatic condition and increase socio-economic challenges. The fertile farmland has turned arid, while fishes have migrated away from the sea due to oil spillage. Also, gas flaring is another major contributor to global warming. Nigeria was listed among the 15 other oil-producing countries that have progressively reduced gas flaring. This is probably a long term achievement on climate change, which is opposed to the negative short-term effects for the economic development.

In order to mitigate the effect of global warming, Nigeria must reduce further emissions and adapt to renewable energy to protect its environment and humanity habiting the space.

By mitigation, it means that measures must be taken by the government to reduce the rate and magnitude of climate change caused by human activities. The mitigation options include reduction in burning of the fossil fuel and reduction of greenhouse gases; reduction of deforestation and increase in forestation and afforestation; modification of agricultural practices to reduce emission of greenhouse gases and build up soil carbon and so on.

By adaptation, it means that we should take measures to reduce the adverse impact of global warming on human life and the environment. It means changing the cropping patterns, stopping further development on wetlands, developing crops that are resistant to drought, heat and salt and strengthening public health.

Climate Change is an attributed cause of flooding because when the climate is warmer it results to: Heavy rains; Relative sea level will continue to rise around most shoreline; Extreme sea levels will be experienced more frequently.

Climate change is therefore likely to increase flood risk significantly and progressively over time. At particularly increased risk will be low-lying coastal areas, as sea levels rise and areas not currently prone to fluvial or tidal flooding as more intense rainfall leads to significantly higher risk of flooding from surface runoff and overwhelmed drainage systems.

The major effect of climate change in Nigeria was witnessed in 2002. Between early July and early September 2012, flooding claimed an estimated 137 lives in Nigeria and forced thousands more to relocate, according to Reuters. In addition to the challenges posed by heavy rains, Nigerians had to cope with the release of water from the Lagdo Dam in neighboring Cameroon, which further swelled the Benue River.

Another noticeable effect of Climate change in Nigeria is desertification. The nation has lost 63.83 per cent of her farmlands to desertification, resulting in massive North-South migration of Northern herdsmen and farmers and constant clashes with other Nigerians over land resources in parts of the country.

Desertification can be defined as a phenomenon of impoverishment of the terrestrial ecosystem under the impact of adverse weather and population activities. The progressive deterioration of the fertile land and loss of its productive capacity renders it unsuitable for human and animal habitation. The United Nation defines desertification as the delimitation or destruction of the biological potential of land which can lead to desert-like conditions. It further described desertification as a process that results after a sand storm. Poor rainfall, overgrazing and over utilization of available piece of land are also major causes of desertification.

Nigeria and Desert Encroachment

The Northern part of Nigeria is endowed with a large expanse of arable land that has over the years proved to be very vital for agriculture and economic activities. Sahara desert has been advancing southwards at the rate of 6.0 per cent within the past three decades. Consequently, the country is losing about 350,000 hectares of land yearly to desert encroachment.

Dr. Newton Jibunoh, is the founder of Fight Against Desert Encroachment (FADE), a Non Governmental Organisation that is committed to the Fight Against Desert Encroachment. The 67 year-old agronomist who crossed the desert in 1965 on his way to go to England for his studies, recently led a group of professionals called the Desert Warriors to Agadesh in Niger Republic. The Desert Warrior was drawn from the Banking sector, Lawyers, Journalists among others, as part of their contribution to sensitize Nigerians about the negative impact of Climate Change. Dr Jibunoh observes that beyond the effects of erosion and demographic pressures on land, deforestation is another primary cause of desertification. According to him, a Wood is an important source of fuel for mostly poor populations in the country who do not necessarily realise the consequences of cutting down trees.

It is essential, therefore, to limit the extent of deforestation and to replant trees but such projects are impossible without the support of local communities. He said that the threat of desert encroachment and desertification are assuming frightening dimension especially as it affects the nation's arable land mass. This has become a source of threat to food production; it is equally believed that the hostile impact of climate change in Northern Nigeria poses serious threat to national security and poverty alleviation strategies in the country as those mostly affected are the most vulnerable in the security that dwell in the villages ravaged by this scourge.

It must be noted that Nigeria has in the past 30 years, witnessed a gradual but consistent encroachment of drought and desertification arising, from the disappearance of large body of water and high activities of dry sand in the Northern part of the country. This situation is further aggravated by destruction of arable and fertile land areas through tree felling for energy, bush burning and overgrazing by herdsmen. According to Nigeria's National Meteorological Agency the rainy season in the north has dropped to 120 days from an average of 150 days when compared with the frequency of rain fall 30 years ago. The result of this is the drop in crop yields by 20 per cent. As at today, it is estimated that over 70 million Nigerians have a direct and indirect experience of the negative impact of drought and desertification.'

Experts have estimated that the affected 11 states namely: Adamawa, Borno, Bauchi, Gombe, Katsina, Jigawa, Kano, Kebbi, Sokoto, Yobe and Zamfara are under intense pressure from the attack of climate change. No fewer than 42 million people are believed to have been affected by this development. We have witnessed the gradual disappearance of fertile lands and steady decline in food production. Lake Chad, the largest body of Inland Freshwater is also heavily devastated by desertification which has come with enormous economic loses and negative impact on the people within the vicinity.

The Buhari administration in 1984, launched a tree planting campaign in Nigeria as a way of bringing the national consciousness to the dangers of rapid depletion of the nation's vegetation and the need to adopt a more sustainable use of forest resources by replanting trees to recover the fading cultivatable lands owing to desert encroachment. It must be observed that the failure in agriculture production, forces people in the affected villages to migrate to more favourable areas, thereby creating new settlements where they have to compete with the indigenous population for the scarce resources. A careful study of the migratory trend in Nigeria shows that there has been a significant displacement of numerous farming and nomadic population in the northern states especially those states that are ravaged by droughts and desertification. Yobe State is one of the worst affected areas in northern Nigeria. Sand dunes are encroaching at a rate of 30 hectares a year, taking over villages. The Yobe State Governor, Alhaji Ibrahim Geidam, has been seeking the intervention of the Federal Government to act fast and save the lives of millions of people in Nigeria. For instance, population of people in eight local government areas are also under severe threat as surface and under-groundwater sources are gradually drying up. The State's Ministry of Environment estimates that more than five million livestock are being threatened by desertification. Other affected areas include Borno, Bauchi, Gombe, Adamawa, Jigawa, Kano, Katsina, Zamfara, Sokoto and Kebbi states.

Gas Flaring and Global Warming

Gas flaring has both local and global impact on the environment. It is one of the most challenging energy and environmental problems facing the world today whether regionally or globally. Flaring of gas is a multi-billion dollar waste, a local environmental catastrophe and a global energy and environmental problem which has persisted for decades.

The World Bank estimated that the annual volume of gas being flared and vented is about 110 billion cubic meters (bcm), which is enough fuel to provide the combined annual natural gas consumption of Germany and France. Gas flaring in Africa (37 bcm in 2000) could produce 200 Terawatt hours (TWh) (200 million MegaWatt) of electricity. (O. Saheed Ismail, G. Ezaina Umukoro, 2012). Gas flaring is a common practice in the oil production process globally. Libya for instance flares about 21 per cent of its natural gas, while Saudi Arabia, Canada and Algeria flare 20 per cent, 8 per cent and 5 per cent, respectively. This implies that Nigeria has one of the worst rates of gas flaring in the world. In 2002, Nigeria flared about 76 per cent of its natural gas.

It is commonly agreed that gas flaring contributes considerably to greenhouse gas (GHG) emissions and has adverse impacts on the environment. The environmental complications caused by flaring are mainly global, but to some extent also regional and local. For example, flaring during oil production operations emits CO_2, methane and other forms of gases which contribute to global warming causing climate change. This negates commitments made by countries under the United Nations Framework Convention on Climate Change (UNFCCC) and Kyoto Protocol. Burning of associated or solution gas produces carbon dioxide (CO_2) and methane (CH_4). These emissions increase the concentration of GHG in the atmosphere, which in turn contributes to global warming. Of these two, methane is actually more harmful than carbon dioxide. It is also more prevalent in flares that burn at lower efficiency. Those less efficient flares tend to have more moisture and particles in them that reflect heat and are said to have similar effect on the ozone layer like aerosols do. Of the greenhouse gases researched so far, the global warming potential of a kilogram of methane is estimated to be twenty-one times that of a kilogram of carbon dioxide when the effects are considered over one hundred years.

The Kyoto Protocol is an international agreement linked to the UNFCCC, which **commits** its Parties by setting internationally binding emission reduction targets.

The Kyoto Protocol was adopted in Kyoto, Japan, on 11 December 1997 and entered into force on 16 February 2005. The detailed rules for the implementation of the Protocol were adopted at COP 7 in Marrakesh, Morocco, in 2001, and are referred to as the "Marrakesh Accords." Its first commitment period started in 2008 and ended in 2012.

Gas Flaring in Nigeria

Nigeria has an estimated 180 billion cubic feet of proven natural gas, making it the ninth largest concentration in the world. Due to unsustainable exploration practices coupled with the lack of gas utilization infrastructure in Nigeria, the country flares 75 per cent of the gas it produces and re-injects only 12 per cent to enhance oil recovery. It is estimated that about two billion standard cubic feet of gas is currently being flared in Nigeria, the highest in any member-nation of the Organisation of Petroleum Exporting Countries (OPEC). This is an enormous flare amount. Consequently, and going by the current statistics, Nigeria accounts for about 19 per cent of the total amount of gas flared globally.

Gas flaring in Nigeria started since the inception of petroleum industry in the country. The rate of gas flaring in Nigeria could make an ignorant person believe that natural gas, unlike crude oil, is not profitable and at best can be treated as a by-product of oil. The Nigerian energy industry was therefore for a long period, inundated with the practice of flaring gas with its environmental and ecological consequences and attendant waste of resources. Nigeria for a long time has remained one of the major gas flaring nations. However, in recent times, the flares are beginning to reduce. The reduction in gas flaring is however due to certain policy interventions put in place by the government to utilize gas.

Figure 1.1: Gas Flaring in Niger Delta Region of Nigeria.

Nigeria has an obligation under the Kyoto Protocols to co-operate with other nations to protect and conserve the environment and reduce its contribution to global warming. Nigeria contributes in several ways to global environmental degradation through large scale deforestation, marine pollution and gas flaring. Nigeria's gas flaring records shows that the country for a long time has been contributing significant amount of greenhouse gases that deplete the ozone layer. Gas flaring in Nigeria therefore affects not only Nigeria but other countries of the world because there are no physical boundaries that demarcate the ozone layer and atmospheric elements are not static. Nigeria therefore shares a common responsibility with other nations to keep the global climate safe for future generations.

Though the role of multinational oil companies in the exploration for crude oil in the Niger-Delta region of Nigeria has remained contentious as a result of the activity of gas flaring that is perpetuated by these companies, it must be pointed out that these companies have made some efforts at reducing the amount of gas flared. However, these reductions do not go far enough in terms of eliminating this unacceptable practice.

Table 1.1: Annual Gas Flaring Volumes for Nigeria over 15 Years (million cubic metres)

Year	Gas Produced	Gas Flared	Per cent of Gas Flared
1996	35450.00	26590.00	75.01
1997	37150.00	24234.00	65.23
1998	37039.00	23632.00	63.80
1999	43636.00	22362.00	51.25
2000	42732.00	24255.00	56.76
2001	52453.00	26759.00	51.02
2002	48192.45	24835.58	51.53
2003	51766.03	23943.03	46.25
2004	58963.61	25090.91	42.55
2005	59284.97	23002.71	38.80
2006	82036.86	28584.39	34.84
2007	84707.34	27307.13	32.24
2008	80603.61	21811.00	27.06
2009	64882.86	17987.59	27.72
2010	67757.65	16468.18	24.30

Source: NNPC (1997); NNPC (2009a); NNPC (2010) Adapted from [62].

Efforts to Curb Gas Flaring in Nigeria

The Federal Government of Nigeria realized the need to curb gas flaring and has since been making efforts to reduce and subsequently stop this wastage. The former Military Head of State, General Yakubu Gowon ordered oil companies operating in the Niger Delta region of Nigeria to work towards ending gas flaring by 1974. This order could not be realized due to the inability of the oil firms to put in place gas utilisation amenities, compelling the government to shift the deadline to 1979. Indications that the country's vision of effective use of its gas resources may ensure long maturation period emerged when the multinational oil firms also failed to meet the 1979 deadline, thus forcing the administration of Alhaji Shehu Shagari to extend the zero-gas flaring deadline to 1984.

To ensure the realization of this objective, an Associated Gas Re-Injection Act of 1979 No. 99 was enacted, this requires oil companies operating in Nigeria to produce detailed plans for gas exploitation as well as guarantee zero flares by January 1, 1984. However, if there are reasons for otherwise, exemption permission should be obtained from the relevant ministry. Although, routine gas flaring was forbidden since 1984, according to Section 3 of Nigeria's Associated Gas Reinjection Act 1979, the practice continued during the succeeding military regimes.

Instead of the much-anticipated reduction, data from Directorate of Petroleum Resources (DPR) show that the rate of gas flaring grew in leaps and bounds owing to the failure of government to enforce the gas flaring law. Besides the zero-gas flaring deadline, the Federal Government also had a number of other regulatory commitments that ought to have helped to realise the objective.

For instance, the National Gas Policy (NGP) first reviewed in 1995 by the late General Sani Abacha regime required subsequent production sharing contracts (PSCs) signed with oil companies to include gas utilization clauses. Incentives were also offered under the Associated Gas Utilisation Fiscal incentives as an effort to put in place investment required to transport gas to interested third parties, yet those measures failed to lead to actualisation of the zero-gas flaring target.

However, many Nigerians heaved a sigh of relief when former President Olusegun Obasanjo's administration kick-started a new gas flaring phase-out in 2000, setting December 2003 as the new deadline for gas flaring phase-out. This followed its renewed investment in the Nigeria Liquefied Natural Gas (NLNG). But the oil firms preferred 2006 as the most realistic date to end the flares.

Although both parties later reached an agreement to end gas flaring by the end of 2004, the Presidency later pushed the date further by two years (2006). However, when the 2006 zero-gas flaring deadline failed to materialise, a new date of 2008 was quickly agreed. While bowing to mounting local and international pressure, government again pledged to halt gas flares in Nigeria by January 1, 2008 as the new zero flare date. It also threatened punitive action for any breach. Again, on December 17, 2007 yet another shift was announced, this time with a deadline fixed for December 31, 2008.

In 2009, the Senate passed the Gas Flaring Bill, making it illegal for operators to flare gas in Nigeria beyond December 31, 2010. Even this deadline was not met, forcing the House of Representatives to propose December 2012 as the new zero-

Figure 1.2: Gas Flaring in Niger Delta Region of Nigeria.

gas flaring date, as well as impose a fine of $500,000 on any company that fails to report, within 24 hours, any emergency flaring on account of equipment failure.

In a highly promising development, the Government of Nigeria has endorsed the global Initiative to end routine gas flaring by 2030 and to disallow routine flaring in new oil field developments onwards.

The Government has set an even more ambitious target. In a letter to the World Bank, Dr. Emmanuel Ibe Kachikwu, Minister of State for Petroleum Resources, says "Nigeria fully supports the global initiative to eliminate routine flaring by 2030; however, we are strategically pursuing a National Target Date of 2020."

Nigeria has already made substantial progress reducing gas flaring and the government remains focused on tackling climate change by further developing the country's gas sector with updated laws and regulations. Proper development of the gas sector remains key to improving the country's electricity sector, where today many gas-fired power plants lay idle for lack of gas availability.

The Bank-led Global Gas Flaring Reduction Partnership (GGFR) will continue supporting Nigeria's Ministry of Petroleum Resources in its efforts to curb gas flaring.

The "Zero Routine Flaring by 2030" initiative, introduced by the World Bank, brings together governments, oil companies, and development institutions who recognize the flaring situation described above is unsustainable from a resource management and environmental perspective, and who agree to cooperate to eliminate routine flaring no later than 2030.

Effect of Climate Change on Food Production in Nigeria

Crops Production: In Nigeria different types of food crops are planted annually. Nigerian farmers cultivate cassava, melon, yam, rice, groundnuts, peppers, onion, plantain, vegetables, *etc.* The cash crops are: cocoa, oil palm, cashew, mango coconut, rubber, cotton and other fruit crops like pineapple, guava, pawpaw, *etc.* All the above crops depend on rainfall. Where rain is abundant especially in the southern parts of the country, crops that require much rain are planted and in the northern part of the country, crops that do not require much rain are cultivated. Food production on the whole has not kept pace with Nigeria's population increases.

Climate change affects crop production in a number of ways, for example, uncertainties and variation in the pattern of rainfall, floods and devastated farmlands, migration pest in response to climate change while high temperatures smother crops.

Irregular and unpredictable rainfall and sunshine hours continue to take the toll on hitherto low-level harvests of rice, maize, cassava, melon, sorghum and yam with at least 2.5 per cent decline of harvests per annum. Cocoa, cashew, oranges, kola nut, oil palm, rubber, cotton and coffee production suffer severe setbacks under reduced photoperiods with flower and fruit abortion trends that shot down annual yields by 5.5 metric tonnes/ha. Pest and disease incidences which become varied and uncontrollable under extreme weather events continue to cause decline in crop harvests, especially that of cowpea, tomatoes, pepper and groundnuts. Drought and flood extremes feature prominently north wards of the country, affecting crops farming and harvests as well as livestock production, the feed of which are mostly crop-based. Flooded farm lands/wetlands expansion cause arable land losses for crops within the areas with limited crop facility capacities and thus reduce root/ tuber crops harvest (yam, cassava, sweet potatoes, Irish potato and cocoyam) by at least 0.25 million metric tonnes per annum.

Flood and Climate Change

Floods can be defined as extremely high flows of river, whereby water overwhelms flood plains or terrains outside the water-confined major river channels. Flood disaster is measured by possibility of occurrence of their damaging consequences, conceived generally as flood risk.

Flooding is one of the major environmental disaster we need to contend with globally. This is especially the case in most coastal areas of the world. Periodic floods occur on many rivers, forming a surrounding region known as flood plain. Rivers overflow for reasons like excess rainfall. The good thing about river overflows is the fact that as flood waters flow into the banks, sand, silt and debris are deposited into the surrounding land.

Increase in global temperatures has contributed immensely to increase in extreme rainfall events. As the Earth's climate warms up, the air is able to hold more water vapor, allowing for more intense downpours that can lead to flooding. Observations have shown that the trends has been on the increase within the past

Figure 1.3: A Village in Bayesa State during the 2012 Flooding.

30 years and the problem will likely grow worse in the coming years, leaving more and more communities vulnerable to the immediate and lingering health impacts of flood events.

In Nigeria, there exist reports of flooding in some towns and cities during heavy downpours. Flood waters from Cameroun entered Nigeria through the Benue River, into the River Niger on its way to the sea. Lots of physical damages were recorded, including destruction of farmlands and houses. Economic life was halted, people displaced and some lost their lives. Although Cameroun released water from the Lagdo dam between July 2nd and September 17th 2012, the waters remained in the Niger delta communities up till November 2012.

Flash floods are common in Nigeria in the rainy season (May to September), but news reports characterized the 2012 floods as the worst in more than 40 years. According to the National Emergency Management Agency (NEMA), 30 of Nigeria's 36 states were affected by the floods. The floods affected an estimated total of seven million people. The estimated damages and losses caused by the floods were worth 2.6 trillion naira

The 2012 rainy season in Nigeria has been worse than earlier years, and heavy rains at the end of August and the beginning of September led to serious floods

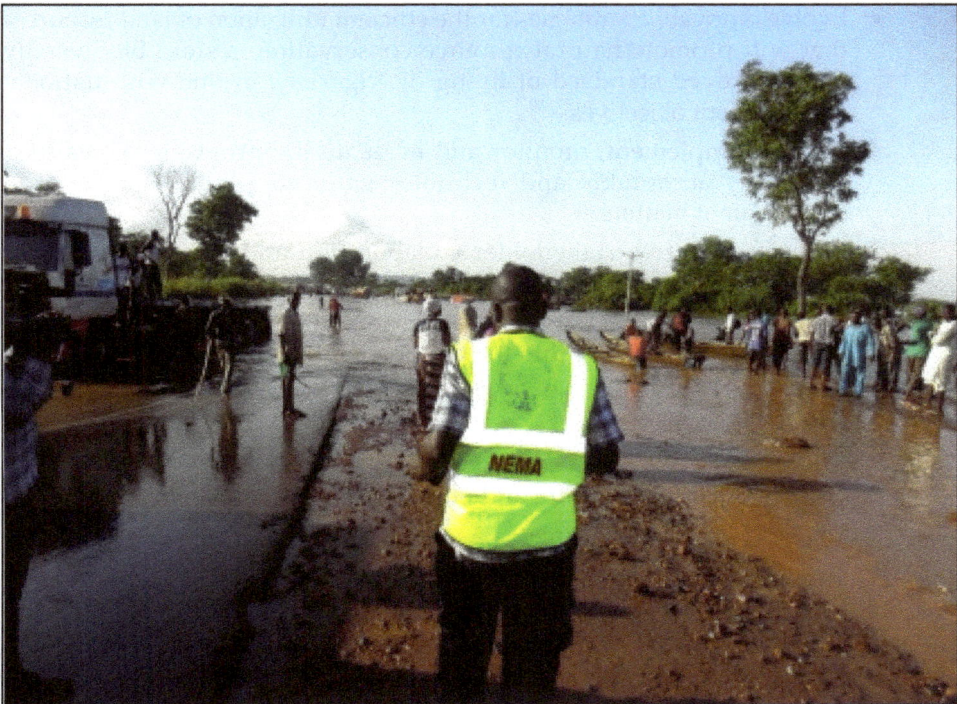

Figure 1.4: Flood on the Ever-Busy Abuja-Lokoja Highway where for Close to One Week, in 2012, the Key Highway Linking Northern Nigeria to the Rest of the Southern Part of the Country Submerged under Water Causing Traffic Delays that have Lasted Days.

in most parts of the country. The Nigerian authorities contained the initial excess run-off through contingency measures, but during the last week of September water reservoirs have overflown and authorities were obliged to open dams to relive pressure in both Nigeria and neighboring Cameroon and Niger, leading to destroyed river banks and infrastructure, loss of property and livestock and flash floods in many areas.

Agencies of the Federal Government of Nigeria Charged with Responsibilities of Protecting the Environment

The Federal Government of Nigeria realised the need to generate information and protect the environment set up some bodies to monitor the environment, provide necessary information about the environment and enforce some of the laws. Some of these bodies and their mandates are as given below:

Federal Ministry of Environment

Mandates

* Formulate and implement relevant policies and programmes that will guarantee that the economic development efforts are in harmony with the environment.

* Evolve innovative strategies for the efficient utilization of land resources that will promote natural resource conservation, restore biodiversity and improved standard of living of Nigerians in line with national development objectives.

* Prepare, implement, monitor and evaluate the provision of realistic, culturally acceptable and technologically adaptable standards in environment matters.

* Promote effective stakeholder collaboration and adequate resource mobilization in programme execution.

* Promote technical cooperation and partnership with bilateral and multilateral agencies within the purview of the ministry's mandate.

Projects/Activities

* Climate Change/Clean Development Mechanism Projects.
* Control of Water Hyacinth and other Invasive, Alien Species.
* Desert to Food Programme.
* Drought and Desertification Control
* Enabling Activities for the implementation of Stockholm Convention and sound management of POPs.
* Erosion, Flood and Coastal Zone Management Projects.
* Greenwall Sahara Nigeria Programme.
* Gulf of Guinea Large Marine Ecosystem Programme.
* Integrated Municipal Solid Waste Management Scheme.

★ Less burnt for a clean earth: minimization of dioxin emission from open burning sources in Nigeria, also known as UPOPs Project.

★ National Afforestation Programme.

★ National Framework for Polychlorinated Biphenyl (PCB) Management Project.

★ National Integrated Food Sanitation Programme

★ Nigeria-Chemical Information Exchange Network Programme (htpp:www.estis.net/sites/cien_ng).

★ Oil Spill Management and Impact Sites Remediation

★ Ozone Depleting Substances (ODS) Phase-out Programme.

★ Reforestation of Degraded Forest Reserve.

★ Regional Project (Ghana and Nigeria) to develop appropriate strategies for identifying sites contaminated by chemicals listed in annexes A, B and C of the Stockholm Convention

★ Sand Dune Stabilization

National Oil Spill Detection and Response Agency (NOSDRA)

NOSDRA was established by the National Assembly of the Federal Republic of Nigeria Act of 2006. It was established with responsibility for preparedness, detection, and response to oil spillages in Nigeria. Its Head office is at 5th floor NAIC House plot 590, zone AO, Central Business District, Abuja. With its zonal offices in Lagos, Akure, Porth-court, Delta, Kaduna, Akwa-Ibom and Bayelsa.

NOSDRA was established in 2006 as an institutional framework to co-ordinate the implementation of the National Oil Spill Contingency Plan (NOSCP) for Nigeria in accordance with the International Convention on Oil Pollution Preparedness, Response and Cooperation (OPRC 90) to which Nigeria is a signatory. Since its establishment, the Agency has been intensely occupied with ensuring compliance with environment legislation in the Nigerian Petroleum Sector. The Agency embarks on Joint Investigation Visits, ensures the remediation of impacted sites and monitors oil spill drill exercises and facilities inspection. It has set up Zonal Offices in Port Harcourt, Warri, and Uyo all in the Niger-Delta region where much of oil exploration and production in Nigeria is carried out and there are also zonal offices in Lagos, Kaduna and Akure. NOSDRA is currently liaising with relevant stakeholders in the Nigerian Oil and Gas Industry to evolve practical methods of environmental management to cope with the dynamics of the Petroleum Sector.

National Environmental Standards and Regulations Enforcement Agency (NASREA)

NESREA, an Agency of the Federal Ministry of Environment is charged with the responsibility of enforcing environmental Laws, regulations and standard in deterring people, industries and organization from polluting and degrading the environment.

Mandates

★ Enforce compliance with laws, guidelines, policies and standards on environmental matters;

★ Coordinate and liaise with, stakeholders, within and outside Nigeria on matters of environmental standards, regulations and enforcement;

★ Enforce compliance with the provisions of international agreements, protocols, conventions and treaties on the environment including climate change, biodiversity conservation, desertification, forestry, oil and gas, chemicals, hazardous wastes, ozone depletion, marine and wild life, pollution, sanitation and such other environmental agreements as may from time to time come into force;

★ Enforce compliance with policies, standards, legislation and guidelines on water quality, Environmental Health and Sanitation, including pollution abatement;

★ Enforce compliance with guidelines, and legislation on sustainable management of the ecosystem, biodiversity conservation and the development of Nigeria's natural resources;

★ Enforce compliance with any legislation on sound chemical management, safe use of pesticides and disposal of spent packages thereof;

★ Enforce compliance with regulations on the importation, exportation, production, distribution, storage, sale, use, handling and disposal of hazardous chemicals and waste, other than in the oil and gas sector;

★ Enforce through compliance monitoring, the environmental regulations and standards on noise, air, land, seas, oceans and other water bodies other than in the oil and gas sector;

★ Ensure that environmental projects funded by donor organizations and external support agencies adhere to regulations in environmental safety and protection;

★ Enforce environmental control measures through registration, licensing and permitting Systems other than in the oil and gas sector;

★ Conduct environmental audit and establish data bank on regulatory and enforcement mechanisms of environmental standards other than in the oil and gas sector;

★ Create public awareness and provide environmental education on sustainable environmental management, promote private sector compliance with environmental regulations other than in the oil and gas sector and publish general scientific or other data resulting from the performance of its functions; and

★ Carry out such activities as are necessary or expedient for the performance of its functions.

Nigerian Meteorological Agency (NIMET)

NIMET came into existence by an Act of the National Assembly – NIMET (Establishment) ACT 2003, enacted on 21st May 2003, and became effective on 19th June 2003 following Presidential assent.

It is a Federal Government agency charged with the responsibility to advise the Federal Government on all aspects of meteorology; project, prepare and interpret government policy in the field of meteorology; and to issue weather (and climate) forecasts for the safe operations of aircrafts, ocean going vessels and oil rigs.

The Act also makes it the responsibility of the Agency to observe, collate, collect, process and disseminate all meteorological data and information within and outside; co-ordinate research activities among staff, and publish scientific papers in the various branches of meteorology in support of sustainable socio-economic activities in Nigeria.

Conclusions

The impact of the climate change will be difficult to handle and it will be potentially very long lasting. The evidence of global warming is increasing daily, and there are risks over and above those that are usually considered. Climate change often appears very obscure but in Nigeria, it's real. We already have an increasing incidence of drought, flood, declining agricultural productivity, and a rising number of heat waves. There is glaring evidence that climate change is not only happening but it's changing our lives. Declining rainfall in already desert-prone areas in northern Nigeria is causing increasing desertification, the states formerly referred to as food basket in central Nigeria are presently not having bumper harvest as before and people in the coastal areas who used to depend on fishing have seen their livelihoods destroyed by the rising waters. Adapting to climate variability and mitigating its impacts is something that we do in our everyday lives, but we have to understand what climate change is, that we contribute to it, and how we can adapt and reduce our vulnerabilities.

An urgent attention or something needs to be done about global warming and climate change. First, there is need to suggest a mechanism for tackling climate change and global warming, the idea of using Carbon Sinks to soak up carbon dioxide from the atmosphere readily comes to mind. Reforestation or planting of new forests, this is a popular strategy for the logging industry and nations with large forests interests like Nigeria. Women and children are particularly the most vulnerable to the impacts of climate change. Inadequate funds hamper progress in achieving Nigeria's objectives on climate change. The Nigerian Government and all the stakeholders involves in the global phenomenon needs to increase public awareness, promote research and establish a commission or an agency that will handle issues related to global warming and climate change. The Federal, State and Local Government, International agencies and other development partners are required to fund climate change projects in Nigeria for sustainable solution.

References

1. O. Saheed Ismail and G. Ezaina Umukoro, "Global Impact of Gas Flaring" Energy and Power Engineering, 2012.

2. Abikoye Oluwatoyosi O, Adhekpukoli Elo S. and Babade Ayobayo J. Nigerian Environmental Law and the Menace of Gas Flaring, 2014

3. Chijoke Evoh, "Gas Flares, Oil Companies and Politics In Nigeria", 2002

4. Olajide Olaniyan, Global warming and effects on Nigeria, 2016

5. http://www.nimet.gov.ng

6. http://environment.gov.ng

Chapter 2

Evolving Energy Models in Emerging Economies Post COP 21: A Perspective of Policy Making on Energy Improvement in Nigeria

Itoandon Emoleila Ejiya[1] and Samuel A. Bankole[2]

[1]Department of Biotechnology,
Federal Institute of Industrial Research, Lagos State, Nigeria
E-mail: emoleila.itoadon@fiiro.gov.ng
[2]Department of Microbiology,
Olabisi Onabanjo University, Ago – Iwoye,
Ogun State, Nigeria

Abstract

Development economics involves the creation of theories and methods that aid in the determination of policies and practices and can be implemented at either the domestic or international level. This may involve restructuring market incentives or using mathematical methods such as inter - temporal optimization for project analysis, or it may involve a mixture of quantitative and qualitative methods. Over the last two decades, the rise of emerging markets and developing economies has transformed the global economy, and in the process, different energy models have reshaped the world's system. With annual growth rates reaching 7.6 per cent during the 2000s, emerging economies grew nine times faster than advanced economies between 2007 and 2014; today, they account for 57 per cent of global output. With 90 per cent of net energy demand growth until 2035 expected to come from emerging economies, understanding the key trends shaping the energy landscape of these economies offers valuable insights for the future of the global energy system. Today, innovation performance is a crucial determinant of competitiveness and national progress. Moreover, innovation is important to help address global challenges, such as climate change and sustainable development but despite the importance of innovation, many developing countries face difficulties in strengthening performance in this area.

Keywords: Energy models, Emerging economies, Policy making, Nigeria.

Introduction

Indeed, many such countries like Nigeria have seen little improvement in productivity performance in recent years despite the new opportunities offered by globalization and new technologies, especially the Information and Communication Technologies (ICT) and mostly in energy. Implementation of policies can support innovation by continually reforming and updating the regulatory and institutional framework within which innovative activity takes place. Market-based activities are now recognized by governments, business and development agencies around the world as potential solutions to major sustainable development challenges – reducing poverty, enhancing livelihoods, protecting ecosystems, tackling climate change, and meeting the Millennium Development Goals.

Increasingly, the private sector is recognizing an emerging business opportunity in designing or modifying business models specifically to address such issues. But to date this evolution has been patchy and ad hoc. Many innovative pilot initiatives have succeeded at a micro-level but crucially failed to achieve success when scaled up. Avoiding this trap will require new approaches. There is no simple reform formula; countries will need to find their own right path, drawing on the successes and failures of others, and balancing the trade-offs across the energy triangle. The modern economy is a complex machine, its job is to allocate limited resources and distribute output among a large number of agents—mainly individuals, firms, and governments—allowing for the possibility that each agent's action can directly (or indirectly) affect other agents' actions in relation to the consumer of economic by product. As economies allocate goods and services, they emit measurable signals that suggest there is order driving the complexity. There also seems to be a negative relationship between inflation and the rate of unemployment in the short term. At the other extreme, equity prices seem to be stubbornly unpredictable.

Learning more about the process that generates these stylized facts should help economists and policymakers understand the inner workings of the economy. An economic model is a simplified description of reality, designed to yield hypotheses about economic behavior that can be tested. An important feature of an economic model is that it is necessarily subjective in design because there are no objective measures of economic outcomes. One of the least understood but potentially important trends in the energy field is how emerging economies' development priorities are shaping energy markets. Emerging economies are expected to make up the bulk of growth in demand for energy in the coming decades, with countries outside the Organization for Economic Co-operation and Development (OECD) accounting for 83 per cent of expected growth in energy demand between 2008 and

2035. As the global centers of expansion, these countries will increasingly influence how new energy markets evolve—commercial frameworks, technology sharing and development, regulations, and preferences for fuels and technologies that meet their societies' needs. Many of these countries have integrated new notions of sustainable development—driven by concerns about local pollution, energy security, climate change, and social development—that are likely to bring about energy systems different from U.S. or European models of energy infrastructure and use.

Factors Influencing Emerging Economic Model

The following factors are often used to evaluate the role of rapidly emerging developing economies in energy development trends and recommend how the development programs along with the international donor community, should shift to capitalize on new trends.

Demographic Change

The world's population is expected to reach 8.3 billion by 2030 and 9.3 billion by 2050. Most of this increase will take place in certain developing countries that are in the early stages of their demographic transition and which will see significant increases in the young working-age population of both sexes. In other developing countries and in most developed ones, the demographic transition is already in its most advanced stage. Fertility rates are low, resulting in an ageing population and in a shrinking labour force. In some of these countries, immigration is likely to be the main source of population growth in the future.

Furthermore, education and urbanization are advancing everywhere in the world. The objective of this section is often to show how these long-term demographic trends are likely to affect international trade patterns through their impact on comparative advantage as well as on the level and composition of import demand. One of the implications of different demographic dynamics across countries is that the distribution of world population will continue to shift towards developing and emerging economies.

Investment

The accumulation of physical capital can affect the nature of international trade in a variety of ways. Greater public infrastructure investment can facilitate a country's participation in world markets by, for instance, reducing trade costs and hence increasing supply capacity. Such investment in physical capital can therefore lead to the emergence of "new players" in international trade. Investment in roads, ports and other transport infrastructure can also strengthen regional trade, while investment in ICT infrastructure can enable a larger number of countries to participate in the ever-expanding international trade in services. Over time, depending on the rate of growth of capital accumulation relative to the rate of growth of the labour force, it is possible for investments in infrastructure and non-infrastructure physical capital (such as plant, machinery and equipment) to alter the comparative advantage of a country already widely engaged in international trade. In an economy where factors of production, such as capital, cannot move across countries, investment must be financed by domestic resources.

Technology

Technological differences between countries are an important determinant of income levels and trade. Empirical research has shown that the accumulation of physical and human capital can only partially explain different income levels across countries (Easterly and Levine, 2001; Prescott, 1998) and different trade patterns. The residual is commonly attributed to technological differences between countries, whereby technology is defined as the information or knowledge required for production. Technological progress is undoubtedly the major factor explaining the fast growth in income in the 19th and 20th centuries. Electrification, the telephone, the internal combustion engine and other breakthroughs have dramatically changed the way the world works. Likewise, technological progress will be a major factor in explaining the future patterns of trade and growth.

Energy

Like labour and capital, natural resources are factors of production that serve as inputs in goods and services production. While there is a broad range of natural resources that could be discussed, the focus here will be on energy. Natural resource prices also tend to be volatile. This has an impact on trade by increasing the uncertainty faced by importers and exporters. The contribution of nuclear energy, hydroelectricity and other renewable sources is small but the share of renewables has picked up in the last decade, driven in part by higher energy prices. The uneven geographical distribution of natural resources means that some resource-abundant countries will have market power in trade. This may create a temptation to exploit that market power through the use of export restrictions. By reducing supply of the natural resource in international markets, the world price of the resource rises, creating a terms of-trade gain for the exporting country and a terms-of trade loss for the importing countries.

Transportation Costs

The cost of transporting goods from producers to users affects the volume, direction and pattern of trade. It determines where the line between tradable and non-tradable goods is drawn and shapes which firms are able to participate in trade and how they organize their production internationally. The cost of transportation is in turn influenced by a wide range of fundamental determinants. These include the geographical features of countries, the quantity and quality of the physical infrastructure that support transportation services, the procedures and formalities used to control the movement of goods from one country to another, the extent of competition in the transportation sector, the pace of technological innovation in the sector and the cost of fuel. The characteristics of the products being shipped also affect transportation costs.

Institutions

How do institutions shape international trade relations? And how does trade affect institutions? The key observation in this section is that, in the long run, there exists a dual relationship between these two variables (in the language of economists, they are endogenous). Put simply, institutions shape and are shaped by international commerce. Understanding this relationship can help shed some light on the future of international trade and the multilateral trading system. What are institutions? Economists have developed a notion of institutions that incorporates practices and relationships as well as organizations. As North (1990) explains, "institutions are the rules of the game in a society or, more formally, are the humanly devised constraints that shape human interaction" (North, 1990). In economics, therefore, institutions are the deep frameworks, such as social norms, ordinary laws, political regimes or international treaties, within which policies – including trade policies – are determined and economic exchanges are structured.

Emering Energy Model

Learning curves (also known as experience curves) are based on the empirical observation that in many industries' unit cost declines along with increases in production experience. This finding was documented in the 1930s for labor input for airframe manufacture. Beginning in the mid-1960s, analyses used the experience curve concept to explain the cost behavior and the resultant market share dynamics in competitive industries. In the energy sector, empirical observation has also supported a learning curve approach. For example, endogenous learning is a component in studies focusing on ethanol production in Brazil and in studies focusing on electricity generation, including nuclear, coal, hydropower, wind, and solar photovoltaics. The system dynamics literature describes the use of learning curves in models of strategy dynamics, technological development, industry growth, and investment decision-making. For example, Naim and Towill (1994) present various learning curve formulations and their application in system dynamics models. Fiddaman's work on integrated climate-economy system dynamics models (1997) explicitly accounts for learning-by-doing. Morrison (2005) uses an extension of learning curve theory to show the potential emergence of positive feedbacks in productivity in the context of throughput constraints. Sterman *et al.* (2007) describe a system dynamics model with learning curves that is used to explore "get big fast" strategies under conditions of bounded rationality for actors in the system. Learning curves are also a prominent feature of a system dynamics model of the photovoltaic industry. This model uses system dynamics and Monte Carlo analysis to examine how the value of PV technology varies over time in response to uncertainty and volatility in 11 key input drivers. A single-factor learning curve is commonly used for analyses in the energy sector. Less commonly used are multiple-factor learning curves that incorporate, for example, research and development investment.

Other Factors Involved in Emerging Energy Models

A better understanding of the factors that shape the demand is necessary anticipating future energy demand in view of current transitions is not easy, especially that the drivers of energy demand are poorly understood and usually include primarily such factors, such as GDP, growth and population, while a broader set of social, economic, environmental, technological and political (STEEP) factors should be included. Feedback mechanisms of changes in energy prices are also rarely taken into consideration. Acknowledging that political, social, cultural and institutional elements can have a significant impact on energy demand, but are not always included in models.

Exploring Demand-Side Uncertainties

Energy demand is most often entered as an exogenous variable in energy-economy models or are derived from reductionist utility models. More accurate estimates of energy demand uncertainties are needed, for example to account for the rebound effect from energy efficiency policy instruments, or behavioural and lifestyle changes. Several insights can be gained from investigating how and the extent to which energy demand is and can be decoupled from economic growth through energy efficiency measures and through reduction of energy demand services.

Disaggregating Energy Demand

Aggregate demand measures are deemed insufficient for understanding drivers of energy demand and for influencing energy consumption. Energy intensity is sector specific. Thus sectoral scenarios are needed to help uncover improvement potentials in energy consumption over different time scales.

Focusing on Energy-Services

Modelers need to analyze changes in structure of energy demand services such as heating and transportation due to individual and collective lifestyles changes. Linking energy sources to the energy services they provide can help structure discussion on energy demand around the fundamentals of consumption choice and inform policy using Cross-Impact Balance analysis (CIB) to systematically include the uncertainty of social factors.

Understanding Lifestyle and Preference Changes

People's preferences are not static, but are influenced by social, technological, economic, environmental and political changes that often trigger lifestyle changes. Some of these changes can emerge from grassroots innovations for sustainable consumption. Several (descriptive) scenarios have been developed based on assumptions about lifestyle changes. This is an emerging area based on empirical evidence that actual behaviour deviates from the rationalistic-economic theory from which energy demand projections are usually derived. Several approaches are being explored such as improved stratification of consumers *e.g.* based on their attitude towards technology adoption and vehicle usage intensity, and taking into account behavioural influences in decision-making.

Adding Behavioural Realism to Energy-Economy Models

Further work is needed to better incorporate assumptions about people's behaviours in energy-economy models in order to be relevant for policy-making. For instance, the method of explorative context scenarios can also help to improve the awareness about the uncertainties of the energy demand expectations associated with lifestyle and preference changes.

Conclusions

Major emerging economies should continue to build capacity for resilience by going through with required reforms, including in the energy sector. Those who fail to act risk becoming "trapped in transition". Business as usual is no longer sufficient to regain the fast growth of the 2000s into the next decades, or to avoid growth from slowing down too much. New imperatives also present an opportunity to better prepare for the future by creating new sources of value. This will include increased technological innovation and support for the development of new sources of supply, with emerging economies playing a key role – two-thirds of energy supply investment is expected to take place by 2035. Transitioning to an energy system that provides more opportunities, requires considerable investment and policies supporting the research and development of new technologies (including digital). These reform architects should bear in mind that effective energy policy design is not sufficient; pragmatic implementation is critical for long-term success. The energy reforms implemented under their watch will survive their time in office; effective reform in this context will be based on maintaining the long-sightedness required to think beyond the immediate future.

References

1.   Lee, R. (2003). The Demographic Transition: Three Centuries of Fundamental Change. *Journal of Economic Perspectives*, **17**(4): 167–190

2.   Mohamed M. Naim, M. M. and Denis R. Towill, D. R. (1994) "Establishing a Framework for Effective Materials Logistics Management", *The International Journal of Logistics Management*, **5**(1): 81 – 88

3.   Morrison, T. (2005) 'Staff supervision in social care', Brighton: Pavilion.

4.   *Naito*, T. and *Zhao*, L. (*2009*). Aging, transitional dynamics, and gains from trade. *Journal of Economic Dynamics and Control*, **33**(8): 1531-1542.

5.   Prescott, E. C. (*1998*). Needed: a theory of total factor productivity. *International Economic Review*, **39** (3): 525–551.

6.   Sterman, J. D., Henderson., R., Beinhocker. E. D. and Newman, L. I. (2007). Getting Big Too Fast: Strategic Dynamics with Increasing Returns and Bounded Rationality, **53**(4): 683–696

7.   William, E. and Levine, R. (2001). What have we learned from a decade of empirical research on growth? It's Not Factor Accumulation: Stylized Facts and Growth Models. *World Bank Econ Rev*, **15** (2): 177-219.

8.   Yakita, A. (2012). Different demographic changes and patterns of trade in a Heckscher-Ohlin setting. *Journal of Population Economics*, 25(3): 853-870.

Chapter 3

Strategy to Achieve the National Emission Reduction Target in the Energy Sector

T. Hardianto[1] and E. Hutrindo[2]

[1]Institut Teknologi Bandung, Indonesia
E-mail: toto@termo.pauir.itb.ac.id
[2]Postgraduate Student in Mechanical Engineering,
Institute of Technology Bandung,
Ministry of Energy and Mineral Resources,
Republic of Indonesia

Abstract

Indonesia is strongly committed to fight climate change as attested by reducing the national Greenhouse Gases (GHG) emissions that had triggered global warming, as stipulated in the country's official document of Intended Nationally Determined Contribution (INDC). The document submitted to UNFCCC COP 21 states Indonesia's reduction of GHG emissions by 29 per cent (832 million ton CO_{2-eqv}) by the year 2030 or even 41 per cent (1.192 million ton CO_{2-eqv}), if it gets international assistance. In order to calculate the target, Indonesia recognizes the need for consolidating both methods and data sources to ensure the high degree of accuracy. Therefore, Indonesia's INDC was developed using dynamic system modelling integrating the aspects of energy, economics and environment. Baseline of national GHG emissions was 2,881 million ton CO_{2-eqv} by 2030 from the sectors of Land Use, Land Use Change and Forestry (LULUCF), Industrial Processes and Product Use (IPPU), energy and waste management. In the energy sector, total GHG emissions were estimated to be 1,444 million ton CO_{2-eqv} by 2030, with the target of emissions reduction is 253 million ton CO_{2-eqv} or equivalent to 17.5 per cent of energy sector baseline for fair scenario and 472 million ton CO_{2-eqv} or equivalent to 32.7 per cent of energy sector baseline for ambitious scenario. To achieve the emission reduction target in this sector, Indonesia has set up the strategy of embarking on a mixed energy use policy, with at least 23 per cent coming from new and renewable energy by 2025. Indonesia has also established the development of clean energy sources as a national policy directive. Energy conservation program emphasises on mandatory energy management to the solid energy user, energy conservation partnership and increasing the efficiency of household appliances. Fuel switching from fuel oil to gas for transportation and household is also expected to significantly reduce the emissions.

Keywords: GHG, INDC, Energy sector, Strategy.

Introduction

Addressing climate change issues as part of Sustainable Development Goals is one of national development priorities for the Government of Indonesia. After ratifying the United Nations Framework Convention on Climate Change (UNFCCC) in 1994, Indonesia has been providing continuous support to the UNFCCC negotiating process towards achieving the objective of the Convention as set out in its Article 2, in line with the Convention principles. By announcing its voluntarily commitment in 2009 to a 26 per cent reduction of national Greenhouse Gas (GHG) emissions by 2020 under Business as Usual (BAU) scenario, and up to 41 per cent with international support, Indonesia has taken a lead in providing example in how then the developing countries could support the Convention objectives. This commitment subsequently promulgated in the Presidential Decree No. 61, 2011 concerning the National Action Plan for Reducing Emissions of Greenhouse Gases (the RAN GRK).

COP 20 at Lima (2014) agreed to reiterate its invitation to all Parties to communicate their Intended Nationally Determined Contributions (INDC) well in advance of the twenty-first session of the Conference of the Parties (by the first quarter of 2015 by those Parties ready to do so) in a manner that facilitates the clarity, transparency and understanding of the INDC. As a follow-up of this agreement, taking into account a variety of significant developments in national development and international dynamics, the Indonesian government decided to conduct a review of RAN GRK as well as prepare INDC document. The review is expected to provide a clear overview of GHG emission reduction achievements as well as an evaluation of the strengths and weaknesses of national efforts to address climate change. The process can yield valuable insights and guidance for the development and implementation of future national climate policy (post 2020 scenario). The review process has to provide a solid basis for the formulation of Indonesia's INDC and should be integrated within the context of national development. Government has set some principles for its purpose:

* ⭐ Indonesia's contribution is voluntary, and is based on the Common but Differentiated Responsibility principle, which takes into account a country's capabilities;
* ⭐ The INDC will be developed based on comprehensive science-policy studies which are supported by the latest information and data (without increasing the workload);

★ The INDC must strengthen existing long-term institutional arrangements, which are also beneficial for future implementation;

★ The INDC must support the policy integration process, in particular climate change policies with non-climate change policies. Economic growth, sustainable development and poverty reduction mutually reinforce and support Indonesia's climate change objectives.

Systems dynamics modelling has chosen as the tool to model emission scenarios of Indonesia's INDC, being suitable for modelling complex environments while allowing stakeholders to participate in the modelling process. This participation is crucial to obtaining early feedback, reliable and credible input, and it increases model transparency and trust in the outcomes. In this model, to make emissions projections, economic activities are clustered in four broad sector categories: the land-based sector, the energy sector, industrial processes and product use (IPPU), and the waste sector. The main drivers for GHG emissions as general assumptions are 5-6 per cent economic growth and 1.3 per cent population growth per year. Simulation of the model resulted total GHG emissions under a BAU scenario as to reach 2,881 million ton CO_{2-eqv} by 2030. That equates to a total growth rate of 3.9 per cent per annum between 2015 and 2030. The energy sector is projected to grow the fastest from 2015 onwards and is the largest emitting sector by 2030, accounting for 50 per cent of national emissions. This is needs to get greater attention in the future. Figure 3.1 shows the contribution of the main sectors between 2000 and 2030.

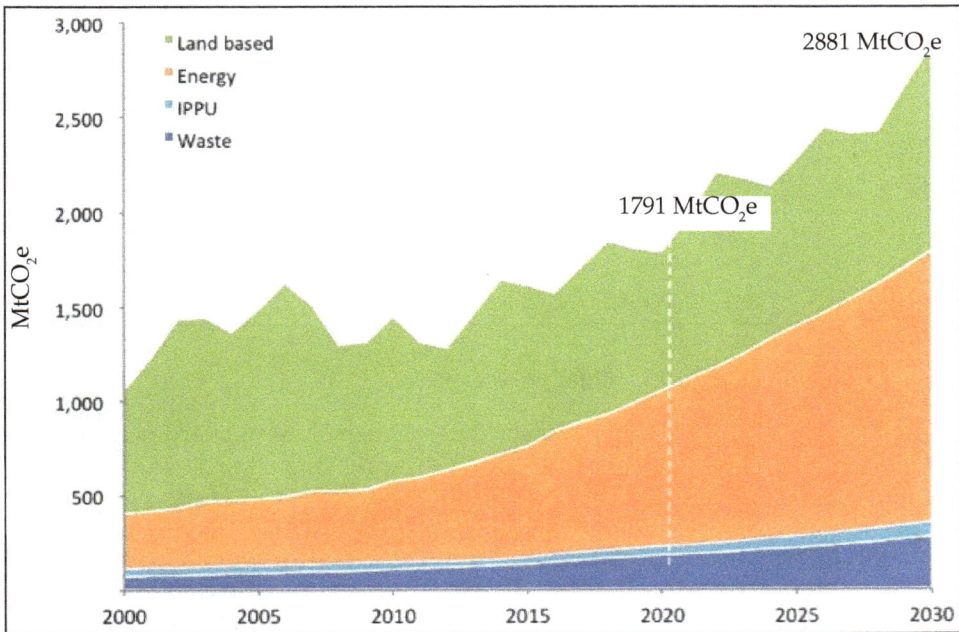

Figure 3.1: Indonesia's GHG Emission Baseline (BAU scenario) by Sector.

There are some mitigating actions simulated in this model for each sector. Considering the potential impacts of existing and future national development policies on GHG emissions, economic, financial, and poverty-reduction, Government has set two mitigation scenarios developed for each sector, called 'Fair' (unconditional) and 'Ambitious' (conditional). Indonesia voluntarily commits to an emission reduction up to 29 per cent relative to the BAU scenario in 2030, equivalent to a reduction of 832 million ton CO_{2-eqv} in 2030. With international support, this target can be extended up to 41 per cent, equivalent to a reduction versus BAU of 1,192 million ton CO_{2-eqv} in 2030. Figure 3.2 shows the two mitigation scenarios modelled for the RAN-GRK review and the INDC and the BAU baseline. The 'Fair' scenario corresponds to the 29 per cent unconditional reduction and the 'Ambitious' scenario corresponds to the 41 per cent reduction conditional on international support.

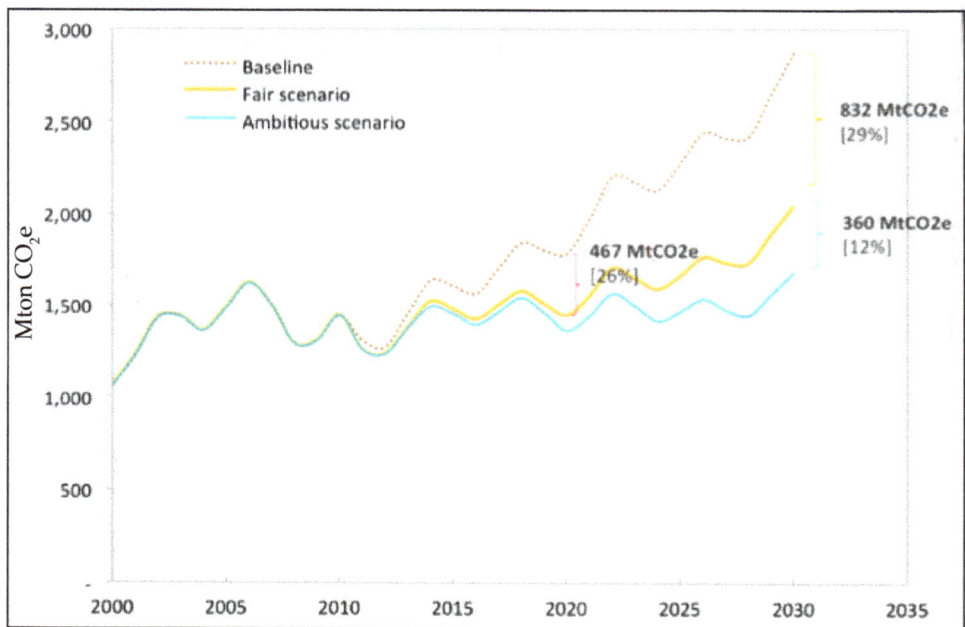

Figure 3.2: Indonesia's INDC Mitigation Scenario.

The BAU baseline for energy sector emissions includes government policies implemented up to 2010. In addition, the target to increase domestic coal consumption by 60 per cent in 2019 is included; consequently the share of coal in the power mix increases. From the latest GHG inventory reported in the Biennial Update Report (BUR), emissions from the energy sector were 458 million ton CO_{2-eqv} in 2010. The RAN-GRK review model projects that BAU emissions from energy will grow at 6.2 per cent per annum from 2015, reaching 1,444 million ton CO_{2-eqv} in 2030. Figure 3.3 shows the projected BAU emissions for the sector and the relative contribution of the different sub-sectors. Emissions from power plants is the fastest growing sub-sector, increasing by 7.3 per cent per annum from 2015 to

2030, reaching 544 million ton CO_{2-eqv}, and accounting for 38 per cent of emissions from the energy sector. Industry and transport are also substantial sources of energy related emissions in Indonesia.

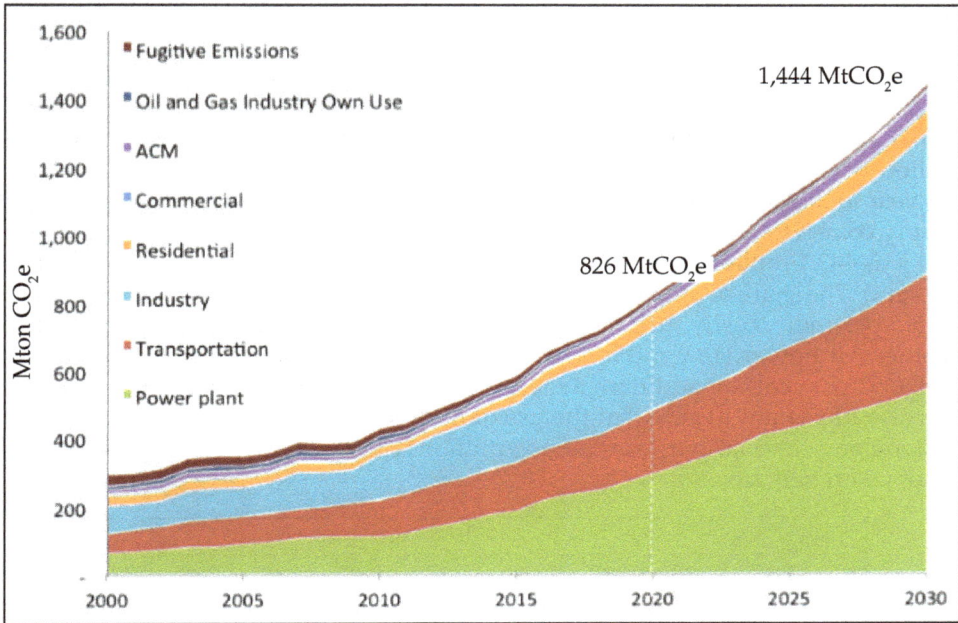

Figure 3.3: Baseline Emissions Projection for the Energy Sector.

Calculation of energy emissions in the model was based on the IPCC 2006 inventory guidelines, which comprise three main subsectors: fuel combustion activities; fugitive emission from fuels; and carbon dioxide transport and storage. The category for carbon dioxide transport and storage was not included as it is still at a very early stage of development. The energy demand modelling included the power plant sector, industrial sector, transportation sector, residential sector, and commercial sector. The projection of energy consumption in each sector is based on assumptions about future economic growth, energy intensity and population growth. Total GHG emissions were estimated to be 1,444 million ton CO_{2-eqv} by 2030, with the target of emissions reduction is 253 million ton CO_{2-eqv} or equivalent to 17.5 per cent of energy sector baseline for fair scenario and 472 million ton CO_{2-eqv} or equivalent to 32.7 per cent of energy sector baseline for ambitious scenario.

Current Status in the Energy Sector

Spread out across more than 17,000 islands, Indonesia is the largest archipelago country in the world. With a population of 247 million, it is also the fourth most populous nation. Sustained high economic growth rates (4.5 per cent - 6.5 per cent annually for the last 10 years) have led to growing demand for basic human needs such as food, housing and water, and energy for household use and productive activities. Due to the interconnected nature of economic, social and environmental factors (including climate change), the effort to provide energy and meet human needs leads to growing GHG emissions and other environmental challenges. Indonesia is also highly vulnerable to the impacts of climate change, and this presents further challenges to its development objectives.

In 2015, Ministry of Energy and Mineral Resources (MEMR) noted that fossil fuels still dominate in the consumption of primary energy (excluding traditional biomass) as shown in the Figure 3.4. Total primary energy supply was about 1,615 million Barrels of Oil Equivalent (BOE), with an average growth of 5.4 per cent per year in the period of last 10 years. The majority of Indonesia's primary energy supply comes from fossil fuels: oil was about 763.5 million BOE (47.26 per cent), coal was 442.1 million BOE (27.37 per cent), and gas was 339.1 million BOE (20.99 per cent). The share of other renewable energy resources in the energy mix was less than 5 per cent, mostly through hydropower (42.6 million BOE or 2.64 per cent), geothermal power (19.9 million BOE or 1.23 per cent), and biofuel (8.2 million BOE 0.51 per cent). In addition, there were less electricity imports from Malaysia. It is also important to note that the use of traditional biomass, prevalent for basic cooking and thermal purposes among millions of rural households in Indonesia, is not taken into account.

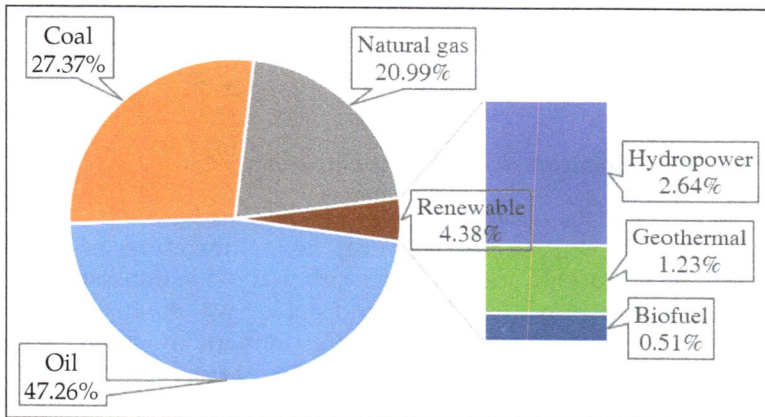

Figure 3.4: Profile of Indonesia Primary Energy Supply 2015.

Consumption of fuel and other derived oil products tend to increase, whereas the developments of oil production over the past 10 years show a downward trend. Indonesia produced 1 million bpd oil in 2005 which decreased to less than 800 thousand bpd in 2015. The decline in production was caused by the ageing of production wells, while the production of new wells is relatively limited. In the period of 2010-2013, Reserve Replacement Ratio (RRR) was about 55 per cent; it means Indonesia produces more oil than the new reserve capacity. Trend of import dependency ratio of fuel and oil also increased from 33 per cent in 2006 to be more than 41 per cent in 2014. This indicates that Indonesia is vulnerable to changes in global conditions; it can even endanger energy security.

In the last ten years, natural gas export (both through pipeline or LNG tanker) was about half of total national production. Or in other words, volume of export is the same as domestic consumption. Total natural gas production in 2015 was 3.1 million MMSCF and almost flat in last decade. Lack of domestic gas utilization is because the price is not competitive yet and the new gas wells being located away from the user/industry there are constrained distribution problems.

Indonesia is the biggest coal exporter in the world, whereas the reserves owned only 3.1 per cent of total world coal reserves. In 2015, Indonesia produced 325 million ton of coal, divided as: 102 million ton for domestic and the rest for export. 83 per cent of total coal export are to China and India. Most of Indonesian coal reserves is medium and low rank coal (HHV < 5.100 kcal/kg). To increase the domestic consumption of coal, it's directed to be fuel in the industrial and the power generation.

Renewable energy is expected to make a substantial contribution in the future. Currently, the utilization of renewable energy resource is limited to hydro, geothermal and biomass used in the power plants. Indonesia produced 1,653 thousand kL of biodiesel from crude palm oil in 2015. 915 thousand kL of biodiesel are used for domestic and blended with the diesel to produce B-20 product (80 per cent of diesel mixed with 20 per cent of biodiesel); it is used as fuel in transportation sector.

Furthermore, final commercial energy consumption in 2015 was reported as 876.6 million BOE (without traditional biomass). This figure has increased by nearly twice since 2000. Share of final energy consumption is divided into: industry (43.5 per cent), household (12.6 per cent), transportation (37.6 per cent), commercial (4.4 per cent), and other sectors (1.9 per cent).

In the end of 2014, installed capacity of power system in Indonesia was about 53.07 GW, consisted of PLN owned 37.37 GW and Non-PLN owned 15.68 GW, with total electricity production reported as 228.554, 90 GWh. National electricity ratio was 87.4 per cent with average national electricity consumption as 952 kWh/capita. If the Indonesian economy continues to grow at its current rate, and considering to increase the electricity consumption per capita as well as electricity ratio, estimates show that domestic demand will rise by around 11 per cent per year, with electricity demand projected to nearly triple between 2010 and 2030.

Other important energy parameter is energy intensity, energy elasticity and energy consumption per capita. Energy intensity was reported as about 2,071 BOE/million USD or equivalent to 290 TOE/million USD in 2015. This energy intensity is still higher than average of OECD countries or even average of ASEAN countries value. Trend of energy intensity in last 15 years indicates decrease although not stable yet, as shown in Figure 3.5(a). Energy elasticity was 0.88, which indicates an improvement in the energy consumption performance, even though the trend line is not stable. Primary energy consumption per capita in 2015 increased by about 10 per cent compared to the previous year. It was about 3.52 BOE/capita and the trend in last 15 years increase is even not stable yet (Figure 3.5b).

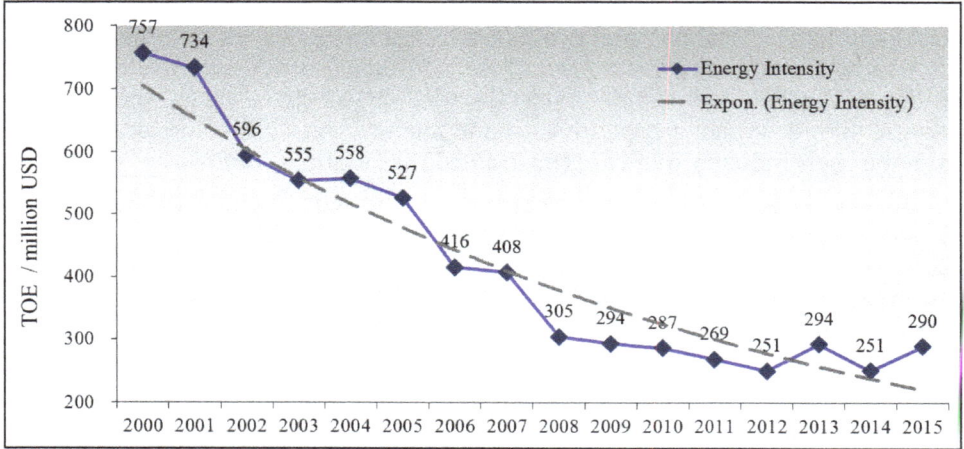

(a) Energy Intensity

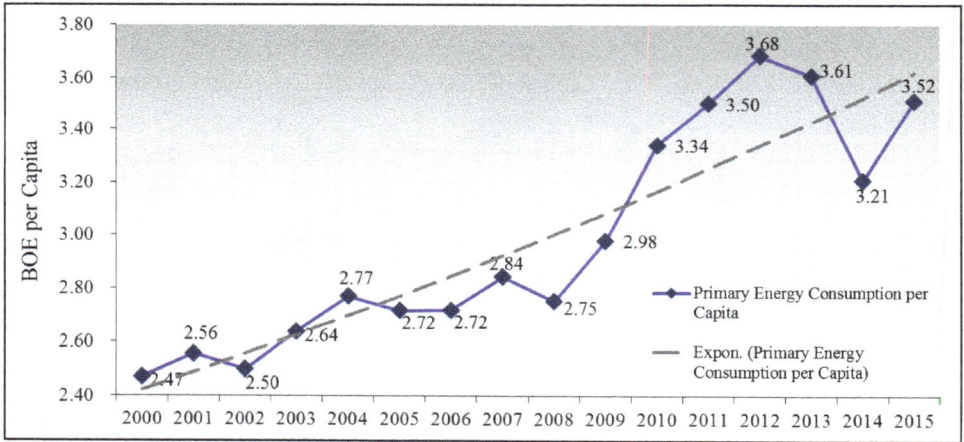

(b) Primary Energy Consumption per Capita

Figure 3.5: Trends of Energy Intensity and Primary Energy Consumption.

Based on energy consumption data that has been described above, total GHG emissions in the energy sector in 2015 was estimated as 507.8 million tons of $CO_{2\text{-eqv}}$. Electricity (power plant) is a sector that emits most GHG emissions, which reached 205.9 million tons of $CO_{2\text{-eqv}}$. Furthermore, the use of energy in industry will push out emissions of 142.9 million tons of $CO_{2\text{-eqv}}$, followed by the transportation sector amounted to 111.7 million tons of $CO_{2\text{-eqv}}$, household and commercial emissions being 27.6 tons of $CO_{2\text{-eqv}}$ and 5.0 tons of $CO_{2\text{-eqv}}$, respectively. Since household and commercial sectors consume a lot of electricity, most of emission for both has been calculated as electricity (power plant) based emissions. Lastly, emission from other sectors was 14.7 $CO_{2\text{-eqv}}$.

Opportunity, Challenge and Strategy

In general, the reduction of GHG emissions in the energy sector can be obtained through the penetration of more efficient technologies and the use of clean energy sources. Both will be involving the potenial/resources characteristic, technology, people and financial/investment aspects. There are opportunity, challenge and strategy to reach the INDC's target of Indonesia emissions reduction, particularly from energy sector.

Opportunity

Indonesia has the potential of renewable energy sources which are clean, vast, variety and considerable. According to MEMR, the country is endowed with significant potential for hydropower (75 GW), solar (4.80 kWh/m^2/day or equivalent to 560 GWp), bioenergy (34 GW), wind (3-6 m/s or equivalent to 107 GW), sea/ocean energy (61 GW), and holds 40 per cent of the world's geothermal reserves (29 GW). Total renewable energy potential in Indonesia was estimated as 866 GW, but only 1 per cent of it has been utilized as shown in the Table 3.1.

Table 3.1: Potential and Utilization of Renewable Energy in Indonesia

Renewable Energy	Potential/Yield	Utilization/Installed
Hydropower	75 GW	5.33 GW
Solar	560 GWp	0.08 GWp
Bioenergy	34 GW	1.74 GW
Wind	107 GW	2.42 MW
Sea/Ocean	61 GW	0.28 MW
Geothermal	29 GW	1.44 GW
Total	**866 GW**	**8.60 GW**

Energy consumption in Indonesia is not efficient yet. This is indicated by the relatively high energy intensity and instability of energy elasticity (sometimes > 1). Study conducted by Directorate of Energy Conservation, MEMR showed that there are energy saving potential in each sector of energy use. Reported that saving energy potential in the industry sector is between 10-30 per cent, transportation sector is 15-35 per cent, household is in the range of 15-30 per cent, commercial buildings is about 10-30 per cent and other sectors is around 25 per cent. High potential saving are in the areas of: air conditioning system, lightings, industrial process and equipment, electrical appliances, transportation and logistics, and smart power grid system. McKinsey study, 2014, reported that if the government energy efficiency optimisation efforts achieve energy saving target of 19 per cent compared to BAU baseline by 2025, it will correspond to save about USD 60 billion in the economy. Furthermore, around 500 energy audit conducted during 2009-2014, in both commercial buildings and industries conclude that energy saving worth about USD 800 million is possible.

Indonesia lies between latitudes 11°S and 6°N, and longitudes 95°E and 141°E, extending 5,120 km (3,181 mi) from east to west and 1,760 km (1,094 mi) from north to south, with tropical climate. Total land area is about 1.9 million km² and sea area is about 3.5 million km². The great opportunity of this is bioenergy development, by optimized land and sea areas. The land area can be used to plant trees or to develop energy forest (chemurgy). Meanwhile in the sea area, cultivation of algae in the marine environment will be promising in the future. Tropical climate also allows the sun shine all year round, it is advantages for metabolism and photosynthesis in plants, as well as potential for solar energy.

Opportunities are there to optimize the utilization of available fossil energy sources in order to reduce emissions. There are some options, *e.g.* shifting fuel in transportation sector from oil to CNG, which is more efficient and less polluting, and increasing gas consumption in the households and power plants. Moreover, Clean Coal Technology (CCT) can be implemented to reduce emissions from coal combustion, particularly at the steam power plant in order to increase the quality of coal (increasing calorific value, reducing moisture, *etc.*) or even high efficient technology to extract energy from coal (*e.g.* ultra super critical steam power plant) can be thought of.

Strong commitment from government as showed in Government Regulation No. 79/2014 on National Energy Policy, regarding energy independence and security to support national sustainable development, government has set energy diversification and energy conservation targets. Energy diversification directed to increase the share of New Renewable Energy (NRE) use in the primary energy mix, which is 23 per cent by 2025 and 31 per cent by 2050. Energy conservation addressed to increase energy efficiency on supply and demand sides, *e.g.* industrial sector, transportation, household and commercial. It is stated that energy intensity is targeted to decrease by 1 per cent per annum till 2025, and energy elasticity less than 1 in 2025.

In 2016, Indonesia has ratified Paris Agreement on Climate Change, in an effort to curb global warming. Thus, Indonesia can actively play role in negotiating process towards achieving the objective of agreement. The real opportunities are: access to global climate fund, joint initiatives (*e.g.* mission innovation, bio future, *etc.*), joint research, promoting clean energy projects, inviting investor or philanthropy to join with climate change project in Indonesia and so on.

Challenge

Energy demand will continue to grow in the future, driven by economic and population growth. Based on the model developed in the preparation of a national energy policy (KEN), primary energy demand is estimated to reach 412 million ton oil equivalent (MTOE) in 2025 or almost twice compared with the year of 2015 (226 MTOE equivalent to 1,615 million BOE). This figure will increase to be about 1,000 MTOE in 2050. The challenge is not only how to find the energy reserves/resources to meet the needs, but also how to optimize the available energy resources, particularly from renewable energy resources, which is abundant, clean, (somewhat) cheaper and sustainable. On the other hand, with fossil energy sources, the challenge is how to conserve energy or consume energy more efficient.

Energy consumption per capita is relatively low (3.52 BOE/capita equivalent to 0.5 TOE/capita). Recent electricity consumption is about 800 kWh/capita, the electrification ratio is 84 per cent, and there are still 12,659 villages not fully electrified yet, even 2,519 villages are totally unelectrified. KEN targeted the primary energy consumption around 1.4 TOE/capita by 2025 and increase to be 3.2 TOE/capita by 2050. Electricity consumption targeted is 2,500 kWh/capita by 2025 and 7,000 kWh/capita by 2050. The challenge is how to increase the production capacity (construction of energy infrastructure) by using reliable and efficient technology with the competitive price/investment. Another challenge is how to electrify the villages. What is the suitable technology and what are the energy resources to be used?

Energy production cost of renewable energy is still relatively high so that it is less competitive compared to the conventional one (fossil resources). The subsidies on fuel and electricity price, make the price of renewable energy even more uncompetitive. financing mechanisms and incentives for renewable energy investors are less conducive. In Indonesia, government has facilitated to create the market of biofuel, subsidized the biofuel and plantation of biofuel feedstock, provided renewable energy feed-in tariff, provided incentives and facilities, provided tax reduction and customs duties, simplified licensing procedures and also provided government special allocated budget for rural energy. However, the development of bioenergy is still relatively slow. The challenge is the breakthrough in the financing, pricing and investment or even in the business model of the renewable energy investment.

Limited availability of information on renewable energy potential is another challenge. Some of potential renewable energy resources lie in areas with difficult accessibility, conserved forest areas, or areas with low energy demand. The challenge is how to arrange or regulate the regional and spatial development. It is important to have specific or thematic map of renewable energy potential in respect to spatial development.

National capacity on renewable energy technology is still limited and dominated by foreign technology. Indonesia adopt the global initiative of improving technology and creating innovations to deploy clean energy. This initiative was announced at COP 21 in Paris and committed by the members to accelerate clean energy research and development efforts, build transparent international collaboration and exchange information required by R&D, build multilateral agreement to reduce or eliminate market barriers for clean energy related goods and services, and build capacity worldwide to harmonize technical standards for provision and maintenance of clean energy systems. The challenge is how to optimize the capacity of national R&D agency/universities to improve technology and create innovations in clean energy technology.

Strategy

In the RAN-GRK review model, strategy of abatement options for the energy sector are based on the policies, considering opportunities and challenges, and measures identified in the Draft of General Planning for National Energy (RUEN),

Table 3.2: Strategy of Emisson Reduction in the Energy Sector

Mitigation Action	Strategy of Abatement		
	Pre 2020 Indicator	Post 2020 Indicator	
		Ambitious Scenario	Fair Scenario
Electricity Supply/Power Plant Sub-sector			
Increase the share of renewable energy in power generation	Based on Electricity Supply Business Plan 2015-2024 share of renewable energy in power generation would reach 9.6 per cent	Based on National Energy General Plan Target/Draft RUEN. Share of Renewable Energy (RE) in power generation: 28.5 per cent in 2025 and 33 per cent in 2030	Based on Electricity Supply Business Plan 2015-2024 share of renewable energy in power generation would reach 16 per cent in 2024
Increase the share of natural gas in power generation	Based on Electricity Supply Business Plan 2015-2024 share of natural gas in power generation would reach 26 per cent in 2019	Based on National Energy General Plan Target/Draft RUEN. Share of natural gas in power generation: 25.5 per cent in 2025 and 25 per cent in 2030	Based on Electricity Supply Business Plan 2015-2024 share of natural gas in power generation would reach 17 per cent in 2025 and 2030
Implementation of Clean Coal Technology (CCT)	Based on Electricity Supply Business Plan 2015-2024 target on implementation of ultra-supercritical coal-fired power plant	Implementation of Ultra-Supercritical (USC) coal-fired power plant would increase the thermal efficiency of coal-fired power plant up to 38 per cent in 2024	Implementation of Ultra-Supercritical (USC) coal-fired power plant would increase the thermal efficiency of coal-fired power plant up to 36.8 per cent in 2024
Industry Sub-sector			
Improve energy efficiency in industry	1 per cent reduction of energy intensity per year from BAU by 2019	1.1 per cent reduction of energy intensity per year from BAU by 2030	1 per cent reduction of energy intensity per year from BAU by 2030
Increase biofuel consumption in industry	Biofuel consumption target by 2019: 0.2 million kL	90 per cent of Draft RUEN Target is achieved. Biofuel consumption target by 2025 : 0.38 million kL	80 per cent of RUEN Target is achieved. Biofuel consumption target by 2025 : 0.38 million kL
Transport Sub-sector			
Improve public transportation	Build electrical train railway with 3 per cent of mode shift	Build electrical train railway with 5 per cent of mode shift	Build electrical train railway with 5 per cent of mode shift
	Build Bus Rapid Transit in 12 cities	90 per cent of Draft RUEN target to build 9550 Bus Rapid Transit in 68 cities is achieved	80 per cent of RUEN target to build 9550 Bus Rapid Transit in 68 cities is achieved

Mitigation Action	Strategy of Abatement		
	Pre 2020 Indicator	Post 2020 Indicator	
		Ambitious Scenario	Fair Scenario
Increase the share of biofuel in transportation sector	Using the blend of up to 20 per cent biofuel in petroleum diesel	90 per cent of Draft RUEN Target is achieved	80 per cent of Draft RUEN Target is achieved.
	Total biofuel consumption target: 4.08 million kL by 2019	Total biofuel consumption target: 7.85 million kL by 2025	Total biofuel consumption target: 7.85 million kL by 2025
Improve the energy efficiency in transport sector by increasing the automobile standard and the implementation of ITS (focus on urban area)	5 per cent reduction in energy intensity reduction from BAU by 2019	5 per cent reduction in energy intensity reduction from BAU by 2030	5 per cent reduction in energy intensity reduction from BAU by 2030
Improve the energy intensity for sea/river transport	0.8 per cent reduction in energy intensity from BAU by 2019	1 per cent reduction in energy intensity from BAU by 2030	0.8 per cent reduction in energy intensity from BAU by 2030
Residential Sub-sector		90 per cent of Draft RUEN Target is achieved	80 per cent of Draft RUEN Target is achieved
Increase the consumption of natural gas in residential sector	Gas demand for residential sector reaches 10.4 MMSCFD	Draft RUEN target: total gas demand for residential sector reaches 440 MMSCFD by 2025	RUEN target: total gas demand for residential sector reaches 440 MMSCFD by 2025
Increase biogas consumption in residential sector	Target of biogas production is 36.865 thousand m^3 in 2019	90 per cent of Draft RUEN target of Producing 22.6 MMSCFD biogas is achieved	80 per cent of Draft RUEN target of Producing 22.6 MMSCFD biogas is achieved
Commercial Sub-sector		90 per cent of Draft RUEN Target is achieved	80 per cent of Draft RUEN Target is achieved
The use of biofuel in commercial sector	Biofuel consumption in commercial sector by 2019: 0.37 million kL	Draft RUEN Target: Biofuel consumption in commercial sector would reach 0.59 million kL by 2025	RUEN Target: Biofuel consumption in commercial sector would reach 0.59 million kL by 2025

Electricity Supply Business Plan (RUPTL) 2015-2024, and Medium Term National Development Plan (RPJMN) 2015-2019. These strategies are presented in Table 3.2.

Conclusions

Total GHG emissions from the energy sector were estimated to be 1,444 million ton $CO_{2\text{-eqv}}$ by 2030 (BAU) or more than triple compared to the base year (2010). By optimizing the opportunities and overcoming the challenges, target of emissions reduction of 253 million ton $CO_{2\text{-eqv}}$ (fair scenario) can be reached with own national efforts. Even additional 219 million ton $CO_{2\text{-eqv}}$ can be reduced if it gets international assistance. So, total emission reduction for the ambitious scenario can reach 472 million ton $CO_{2\text{-eqv}}$ or equivalent to 32.7 per cent of energy sector baseline.

Emissions reduction in the energy sector can be obtained through two main activities: (i) energy efficiency, and (ii) the use of clean energy sources, implemented in each sub sector.

References

1. Climate Change National Board (DNPI), 2010. Indonesia's Greenhouse Gas Abatement Cost Curve. McKinsey Global Institute Analysis. DNPI, Jakarta, Indonesia.

2. Directorate General of New, Renewable Energy and Energy Conservation MEMR, 2016. Renewable Energy Policy: Indonesia. Paper Presented at the Workshop on Renewable Energy Policy Study, Bangkok, Thailand.

3. Government Regulation of Republic Indonesia No. 79 Year 2014 on National Energy Policy (KEN).

4. Ministry of Energy and Mineral Resources, 2016. Handbook of Energy and Economic Statistics of Indonesia 2015. Center for Data and Information Technology MEMR, Jakarta, Indonesia.

5. Ministry of Energy and Mineral Resources, 2015. Statistic of Electricity 2014. Directorate General of Electricity MEMR, Jakarta, Indonesia.

6. Ministry of Energy and Mineral Resources, 2015. Final Draft of General Planning for National Energy (RUEN). Planning Bureau MEMR, Jakarta, Indonesia.

7. Ministry of Environment and Forestry (KLHK), Indonesia Biennial Update Report (BUR). The 3rd National Communications to the UNFCCC. KLHK, Jakarta, Indonesia.

8. Ministry of National Development Planning/BAPPENAS, 2015. Medium Term National Development Plan (RPJMN) 2015–2019. BAPPENAS, Jakarta, Indonesia.

9. Ministry of National Development Planning/BAPPENAS, 2015. Developing Indonesian Climate Mitigation Policy 2020 – 2030 through RAN-GRK Review. BAPPENAS, Jakarta, Indonesia.

10. Ministry of National Development Planning/BAPPENAS, 2015. Dokumen Pendukung Penyusunan INDC Indonesia. BAPPENAS, Jakarta, Indonesia.

11. National Energy Council (DEN), 2016. Indonesia Energy Outlook 2015. Secretariat General of DEN, Jakarta, Indonesia.

12. Patterson, S., 2015. Indonesia's Energy Requirements – Part Two: Future Energy Demands. Associate paper for independent strategic analysis of Australia's Global Interest. Future Directions International, Australia.

13. Presidential Decree of Republic Indonesia No. 61 Year 2011 on National Action Plan to reduce GHG emissions (RAN-GRK).

14. PT. PLN (Persero), 2015. Electricity Supply Business Plan (RUPTL) 2015–2024. Planning Directorate, Jakarta, Indonesia.

15. Tharakan, P., 2015. Summary of Indonesia's Energy Sector Assessment. ADB Papers on Indonesia, Asian Development Bank, Jakarta, Indonesia.

Chapter 4

The Implementation of Energy Development Priorities on the Bases of Efficiency, Saving and Deployment of Renewable Energy in the Province of Havana, Cuba

Osleidys Torres Valdespino

Chief of Science and Technology,
CITMA Delegation of Havana, Cuba
E-mail: adela@delegcha.cu, osleidys@delegcha.cu

Abstract

Energy consumption is valued as an index of a country's economic and social progress. The energy problem today is of crucial importance, not only from the point of view of meeting the growing demand, but also in terms of its environmental and social impact. The irrational way in which fossil fuels have been used has greatly damaged nature. Within the priorities of Cuba and the province of Havana in the energy sphere are the generation and consumption of energy with efficiency, promoting the development of distributed generation and saving measures. It also promotes the use of renewable energy by developing new energy facilities. The paper analyzes the national projection for the country's energy development, where it is projected to reduce the use of fossil fuels to 76 per cent and develop renewable energy sources to 24 per cent by 2030.

Keywords: *Energy development, Renewable sources of energy, Energy efficiency, Projects of development.*

Introduction

The consumption of energy is valued as an index of economic and social progress. The energy problem today is of crucial importance, not only from the point of view of satisfying growing demand, but also in terms of its environmental and social impact. The irrational way in which fossil fuels have been used has greatly damaged nature. The medium- and long-term solution is based on the large-scale use of renewable energy sources.

The Republic of Cuba is located in the archipelago of the Sea of the Antilles. For its political-administrative direction, the territory is organized in fifteen provinces and a special municipality. The province of Havana is its capital and the most populated city. It is the political-administrative center, the seat of legislative power, executive and of the central administration of the State; It hosts most of the scientific-technical activity and specialized services, a considerable part of industrial production (except the sugar industry) as well, the main tourist pole of the country.

The electricity generation has been based on fossil fuel in thermoelectric plants with the installation of generators with a capacity of 2000 MW. They use fuel oil and diesel as basic fuels, which have required a specialized technical force and management at different levels that allows the operation, maintenance, control and inspection of such facilities. Work is on to achieve efficient use of energy, reducing consumption rates through development of a program to switch from high power electrical equipment and electric light bulbs to energy savers.

Also, the use of renewable energy is promoted, through investments on new energy facilities, such as: wind and photovoltaic parks, biogas production and cogeneration with the use of biomass as a way to reach projected share of 24 per cent of generation in 2030.

Materials and Methods

I. The Program of energy development of Cuba has as one of the fundamental support elements, research and innovations to be carried out by the entities of the sector located in the province of Havana, with the 4set goals:

 ⭐ Development of the use of renewable energy.

 ⭐ Introduction of efficient energy management technologies, generating savings in the industrial sector.

 ⭐ Application of technologies to ensure the efficient use of energy in urban transport and freight transport.

 ⭐ Improvement of the electrical energy system.

II. For the elaboration of this paper the following activities were developed:

 ⭐ *Documentary Analysis*: Reports of research and results reported by the Science, Technology and Innovation entities based in Havana, information on innovations applied by companies, government documents and reports on energy priority were collected.

★ Development of interfaces for the introduction of research outputs as reported by ECTIs and their assimilation by companies has also been studied.

★ *Historical - Logical Method*: Evaluation of the actions carried out with the energy priority in the CITMA Delegation of Havana has been done. In this regard, these actions were reviewed in the following ECTI and companies:

Research Centers

1. Electro Energy Research and Testing Center
2. Center for the Study of Renewable Energy Technologies
3. Petroleum Research Center

Technical Scientific Services Centers

1. Center for Information Management and Energy Development
2. Refrigeration and Air Conditioning Institute

Business of Ministry of Energy and Mines

1. "Ñico López" Oil Refinery
2. Maintenance Company for Power Plants
3. CUBALUB
4. Basic Electrical Organization
5. GEYSEL
6. INEL

Results and Discussion

The policy of "energy transition" to reach a sustainable economy with significant use of Renewable Energy Sources (RES), energy efficiency and sustainable development was discussed and approved in the National Assembly of the People's Power of Cuba. There, strategies closely linked to the management and technological innovation developed by professionals and institutions of the country were drawn.

Distinguished within the programs are:

★ Increased efficiency in the use of fossil fuels (non-renewable sources of energy)

★ PAEC- Electricity Saving Program in Cuba

★ Eolic Energy Development

★ Development of Photovoltaic Solar Energy

★ Use of biomass from sugar cane and forest

★ Development of Hydroenergy

★ Biogas Development

Energy Matrix of the Country and its Projection (Ministry of Basic Industry of Cuba, 2014)

The country's energy matrix in 2013 had 96.7 per cent of fossil fuel use, with 48.3 per cent of crude oil, 9.6 per cent of accompanying gas, 4.2 per cent of diesel and 33.6 per cent of fuel oil, 4.3 per cent of renewable energy, where 3.5 per cent of biomass is reported, 0.1 per cent of wind energy, and 0.7 per cent of hydro.

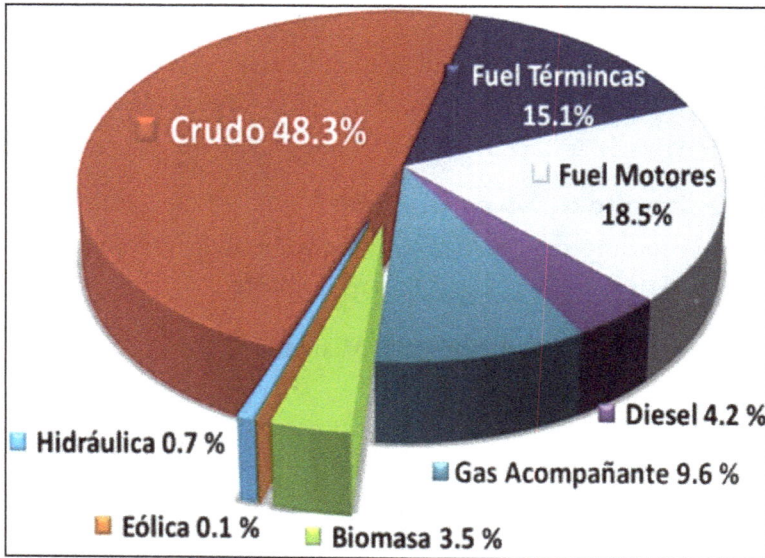

Figure 4.1: Energy Matrix 2013.

In the projection for 2030, it is planned to reduce the use of fossil fuels to 76 per cent, which include: 32 per cent crude oil, 21 per cent liquefied natural gas, 8 per cent accompanying gas, 14 per cent fuel oil and 1 per cent diesel. To increase renewable energies to 24 per cent, through the development of the use of biomass to 14 per cent, solar energy to 3 per cent, wind to 6 per cent and hydropower to 1 per cent.

National System of Electrical Generation (National Bureau of Statistics, 2008)

Cuba has a network of 9 thermoelectric plants distributed in 8 provinces, with a generating capacity of about 2,500 MW, which is below the demand of the country, an aspect that is complemented by the use of 4 units of energy, employment of renewable energies and batteries of emergency generators located in production and service centers.

Electricity Saving Program in Cuba (PAEC) (Ministry of Basic Industry of Cuba, 2008)

In 2006 a transformation of the energy policy, called the Cuban Energy Revolution (REC), was launched in the country, which had 6 national lines of action:

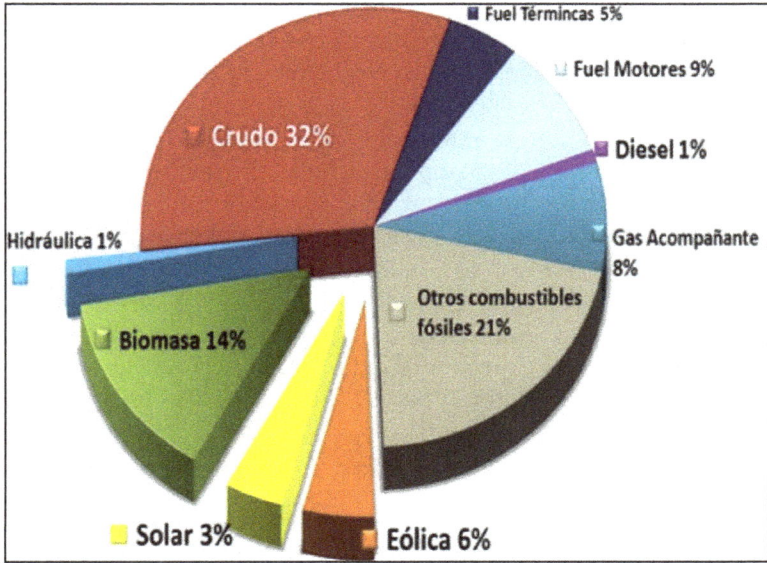

Figure 4.2: Projection of the Energy Matrix for 2030 (Ministry of Basic Industry of Cuba, 2014).

Distributed generation: generation sets were installed with diesel and fuel engines.

1. Energy self-sufficiency: through the use of oil and natural gas, research and oil prospecting and technological improvements.

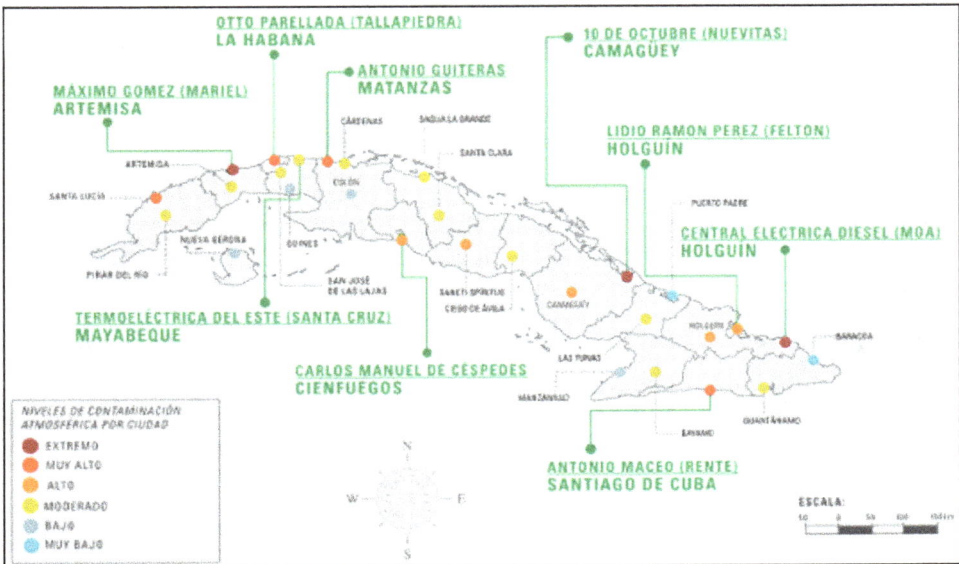

Figure 4.3: Network of Thermoelectric Plants Generating Energy from Fossil Fuel.

2. Rehabilitation in transmission and distribution networks: as a way to reduce losses by replacing posts, connections and circuits, replacing analogue electric meters with electronics in the residential sector and eliminating low-voltage areas.

3. Increased efficiency: massive change of appliances (refrigerators, air conditioning, televisions and fans), water pumps, electric bulbs (incandescent by fluorescent); popularisation of the electric cooking system (burners, multipurpose pots, rice cookers, *etc.*), eliminating the use of kerosene and reduction of Liquefied Petroleum Gas (LPG) and increasing electricity tariffs according to consumption levels.

4. Participation and awareness through the involvement of student brigades and social workers to promote a culture of energy saving, awareness in the media and by the Electricity Saving Program of the Ministry of Education.

5. Development of Renewable Energy Rources (RES): wind, solar, hydro, biomass, as well as national and international cooperation for its development.

The objectives of the program developed in Cuba were:

1. Reduce the maximum demand and the annual growth rate of consumption, so that the demand is lower than that of the gross generation and that it grows less than the Gross Domestic Product.

2. To develop habits and customs in the rational use of energy and protection of the environment in the new generations.

3. Develop a regulatory base and pricing policy that encourages the rational use and management of demand and increases the energy efficiency of all new electrical equipment installed in the country.

4. To date, about 15 million lamps and inefficient electrical appliances have been replaced, including incandescent lamps, fans, water pumping equipment, televisions and food cooking equipment, as well as the development and commercialization of induction cookers.

The average consumption of electric energy in Cuban households has remained practically unchanged after the distribution of equipment such as the kitchen modules, which meant a considerable increase in the quality of life of the population that used kerosene for the cooking of food. (Calvó and Ibáñez, 2009)

Development of Renewable Energies (C. Moreno, 2015)

The use of renewable energy sources is part of the economic and social development programs of the country. Its use is one of the main priorities for the country and indispensable for the achievement of the following objectives:

★ Reduce the inefficiency of the electrical system.

★ Reduce dependence on fossil fuels.

★ Contribute to environmental sustainability.

★ Modify the energy matrix of generation and consumption.

★ Increase the competitiveness of the economy as a whole.

★ Reduce the high cost of energy that is delivered to consumers.

There are currently 30499 energy facilities using the FRE, although not all are properly used, due to difficulties with poor maintenance and operation:

★ 4 wind farms (11.7 MW).

★ 9476 solar panels in isolated installations of the grid.

★ 7 photovoltaic parks connected to the grid (11 MW).

★ 10 595 solar collectors for water heating.

★ 57 sugar producing power plants with sugar production residues.

★ 827 biogas plants.

★ 9343 windmills.

★ 180 hydroelectric installations.

Wind Farms

On the north coast of the island in the eastern zone there are 21 areas where they are the most advantageous for the installation of wind farms. Adding the potential of these 21 areas, the estimated value of the installed capacity in all wind farms that can be located in the good wind zones of the country is from 1,800 to 3290 MW.

At present, there are 4 wind farms installed in the country:

★ Ciego de Ávila, (Turiguano): 2× 225kW

★ Isla de la Juventud, (Los Canarreos): 6 ×275 kW

★ Holguin, (Gibara 1): 6 × 850kW

★ Holguin, (Gibara 2): 6 × 750kW

The technical analysis recommends the installation in the Electrical System of about 633 MW in 13 places of the provinces of Ciego de Avila, Camagüey, Las Tunas, Holguín and in the Maisí municipality of the province of Guantanamo, at a cost of 1120 MMUSD with a recovery time of 4 to 6 years. Work is already under way on the construction of a 51 MW wind farm in the province of Las Tunas. It is strategic the If participation of the national industry in the manufacturing of the components of a Wind Turbine is strategically planned, this would reduce the costs of import and create sources of employment. Once the entire capacity is installed, its estimated generation will be 1630 Million kWh/year, that is, 5.4 per cent of the total energy envisaged for the year 2030.

Windmills

At present there are more than 9343 mills installed throughout the archipelago that are produced in the country and sold in the domestic market, being used mainly for the water pumping in the livestock sector, although all are not working.

Photovoltaic Solar Energy (Off-grid and grid-connected)

Cuba so far has installed about 3 MW of PV, basically isolated systems, solving social needs in remote areas. More than 9000 installations provide the services with a high social impact. According to estimates, with 100 km² of photovoltaic installations can generate 15 000 GWh/year, which equals the current generation based on conventional fuels.

Electrical Cogeneration from Biomass

In Cuba the most abundant biomass is sugar cane for the development of energy through cogeneration, and is currently the only one from which electricity is being generated. There is great potential in the use of Marabú as forest biomass. The sugar industry program aims to implement between 2015 and 2030 not less than 715 MW of Bioelectric plants, attached to 57 sugar mills that will grind between 150 and 180 days per year not less than 85 per cent of its capacity, with a steam consumption of the sugar process of less than 400 kilograms per tonne of ground cane.

Development of Hydro Energy

Currently, in Cuba, there are 180 hydroelectric facilities: 1 major hydroelectric plant, 7 small hydroelectric plants, 35 mini hydroelectric plants and 137 micro hydroelectric plants. The total installed capacity is 62.22 MW, with an electricity production of 149.5 million kWh/year.

Biogas Production

Biogas technology is well adapted to the ecological and economic requirements of the future, and is considered to be state-of-the-art technology. In Cuba, a program has begun to develop biogas installations in the livestock sector, as a way to reduce levels of environmental pollution while ensuring the energy demanded by livestock facilities (dairy and cochiqueras) and organic matter and nutrients to be used in agricultural production of pastures and vegetables. At present, there are 827 biogas plants installed, with a program under development.

Interface Actions Developed for the Control and Execution of the Prioritized Research and Innovation Activities

The CITMA Delegation in Havana is the organization responsible for applying, guiding and controlling the policies drawn up by the State and Government on Science, Technology and Environment to contribute to the sustainable development of the country's capital. To this end, the following actions were carried out:

Exchange of the Work Carried Out by the ECTIs and Analysis of the Projects Carried Out

The Entities of Science, Technology and Innovation that pay for this priority carried out a total of 96 research projects, of which 15 are completed in 2015; these are in the phase of introduction in different companies related to the energy activities.

Table 4.1: Projects Completed in 2015 and under Execution during 2016

	Projects Completed in 2015	*Projects in Execution in 2016*
Center for the Study of Renewable Energy Technologies	-	9
Electro Energy Research and Testing Center	5	19
Refrigeration and Air Conditioning Institute	3	9
Petroleum Research Center	5	48
Center for Information Management and Energy Development	2	11
Total	**15**	**96**

Source: Reports of visits to Entities of Science, Technology and Innovation.

CITMA Delegation of Havana

Follow-up of the National Science, Technology and Innovation Programs linked to Energy Priority

This priority has two National Science, Technology and Innovation Programs coordinated by the Center for Information and Energy Development of the Ministry of Science, Technology and Environment, and the Organisms of the Central Administration of the State related to energy development. Its main objectives are: the efficient use of energy, the use of renewable energy sources, the evaluation and control of environmental pollution and the implementation of clean technologies, as well as the development of the overall general culture of society.

The programs in execution are:

1. **"Sustainable Development of Renewable Energies", (2014-2017)**: In this call, 48 projects were approved, of which 8 projects are subject to the priorities established for Havana.
2. **"Energy Efficiency and Conservation" (2015-2018)**: To date, 11 projects have been received, as most of the projects under development are in the business category. Of these 11 projects, only 2 are executed by entities from the province of Havana.

Evaluation of the Introduction of the Most Relevant Results by Prioritized Goals

Goal I: Development of the Use of Renewable Energy

1. Evaluation of the solar potential for electricity generation with photovoltaic systems connected to the grid in the province of La Habana (Basic Electric Organization and National Electrical Union.
2. Technological development of a solar dryer for agricultural products in Urban Agriculture, municipality of Boyeros (Center for Information Management and Energy Development)

3. Geographic information system for the territorial exploitation of sources of Renewable Energy (Research Center and Electroenergetic Testing)

Goal II: Generalization and Introduction of Efficient Energy Management Technologies, Generating Savings in the Industrial Sector

1. Revision of the Energy Consumption Index Standard elaborated by the PAEC CTN 77 for household refrigeration equipment (Refrigeration and Air Conditioning Institute).

2. Certification of energy consumption of all refrigeration and air conditioning equipment that are imported and manufactured nationally to guarantee the country's energy program according to MINBAS Resolution 136 (Refrigeration and Air Conditioning Institute).

3. Evaluation of Chinese technologies of induction cookers and determination of the technical requirements for their importation and/or production in Cuba (Research Center and Electroenergetic Testing).

4. Creation and Development of Absorption Capacity in Productive Base Organizations of the Cuban Distributed Generation (Research Center and Electroenergetic Testing.

Goal III: Application of Technologies to Ensure the Efficient Use of Energy in Urban Transport and Freight Transport

1. Development of chemometric methods of pattern recognition for the quality control of Jet A-1 turbocombustibles based on their physical properties (Petroleum Research Center).

2. Study of energy indicators in transport equipment (Center for Engineering and Environmental Management of Transport)

3. Improvement of the quality of the asphalts of domestic production through the use of polymeric modifiers (Petroleum Research Center)

4. Improvement of the quality of high octane gasoline (Petroleum Research Center)

Goal IV: Improvement of the Electro-energetic System

1. Maintenance and updating of software developed for the analysis of the National Electroenergetic System that are used daily in the National Dispatch Office (Research Center and Electroenergetic Testing)

2. Procedure that guides the necessary rules regarding the safety of the National Electroenergetic System (Research Center and Electroenergetic Testing)

3. Development of procedures for the restoration of SEN (Research Center and Electroenergetic Testing)

4. Evaluation of alternatives for fuel economy by means of transport (Center for Engineering and Environmental Management of Transport)

Links between ECTI and the Business Sector

It has to ensure that the research developed respond to the demands of business, as a way to achieve the closed cycle of research, highlighting:

* ✶ Research Center and Electroenergetic Testing respond to the research needs of the Electricity Union, guaranteeing research aimed at improving the distribution network, and energy efficiency.

* ✶ Petroleum Research Center works for Union Cuba Oil, with a view to the detection and exploitation of oil and gas fields, as well as the development of additives.

* ✶ Center for Information Management and Energy Development develops a number of tasks included in the prioritized programs carried out in the country in order to respond to the priorities established nationally, mainly directed to the problems of renewable energy.

* ✶ The Institute of Refrigeration and Air Conditioning, advises and provides technical services on the rational use of energy on the topics of Refrigeration and Air Conditioning.

* ✶ Center for the Study of Renewable Energy Technologies develops research and services related to the needs of various agencies in the areas of Renewable Energy.

Conclusions and Recommendations

1. Develop advanced and efficient technologies, prioritizing those that use renewable energy sources.

2. Continue studies of the potential of renewable energies in the country and its feasibility.

3. Conduct studies of energy vulnerability to climate change and other factors in each of the existing technologies - a particular case is biomass.

4. Consult successful experiences of the implementation of renewable energies in tropical conditions.

5. Promote the use of renewable energies in the residential sector, where the energy consumption is high in the country.

6. Encourage and promote technological change in the most energy-intensive sectors, or in key sectors of the economy such as tourism.

References

1. Collective of authors, (2007) Ten questions and answers on wind energy. Editorial CUBASOLAR.

2. Cuba. Ministry of Basic Industry (2008). The Energy Revolution. Concepts and results.

3. Cuba. Ministry of Basic Industry (2014). Energy matrix of the country and its projection.

4. Díaz Canel Bermúdez, M. (2015). XX1 Conference of the Parties to the United Nations Framework Convention on Climate Change. Paris France.

5. Moreno Figueredo, C. (2015). Cuba to 100 per cent with renewable energies.

6. National Bureau of Statistics (2008). Energy statistics in the Revolution.

7. Velázquez León, S. (2016). Energy perspectives in Cuba. Spanish Institute of Strategic Studies. Newsletter.

Chapter 5

An Approach and Priority Analysis on Development Model using Emerging Energy Technologies for Newly Industrialized/Developing Countries

Levent Yagmur

The Scientific and Technological Research Council of Turkey (TÜBİTAK)
Marmara Research Centre (MAM), Energy Institute
Baris Mah. Dr. Zeki Acar Cad. No:1 P.K. 21
41470 Gebze Kocaeli TURKEY
E-mail: levent.yagmur@tubitak.gov.tr

Abstract

The COP 21 is an important milestone for the world since many countries agreed to avoid the rising of global temperature. National and international climate restrictions are one of the main issues for newly industrialized/developing countries for adaptation after COP 21. Developing countries should take care of the new energy parameters in their growth perspective to close the gap between developed countries. Suitable models can provide economic growth. Most of developing countries do not have their own emerging energy technologies. Therefore, technology transfer is an option and its localization is crucial for newly industrialized/developing countries. However, such technologies need to be adapted, being taken from abroad, to the country's local conditions. Countries need to be ready to enhance both its infrastructure and its manpower to be a good competitor in the market. Priority analysis at the beginning can help to choose suitable technologies to save time. In this study, new energy concepts and their parameters are considered to analyse their priorities.

Keywords: *Emerging energy technologies, Technology development, Technology transfer, Localization of technology, Multi-Criteria Decision Making (MCDM), Analytic Hierarchy Process (AHP), Priority analysis, COP 21, Decarbonisation.*

Introduction

The COP 21 was an important milestone since 196 countries agreed that climate change is a great global threat and signed an agreement particularly to keep the rising global temperature below 2°C by mid-century (called "2DS"). This is the first time in history that most of the nations agreed by a consensus to take actions for decarbonisation under a general legally binding framework (IEA, 2016a).

To adopt national and international climate restrictions is one of the main issues for newly industrialized/developing countries. This new situation forces new technologies. These countries must develop suitable model for their sustainable economic growth. For sustainable growth these countries should have their own energy technologies. It is difficult to develop technologies in energy field in short times. Most of developing countries do not have indigenous technologies in energy sector. Therefore, technology transfer and its localization are crucial for newly industrialized/developing countries.

Fast, radical and effective policy actions are required to provide transition to low-carbon energy sector. All actions should be supported by research in applied science and commercialization for deployment to make decarbonisation aided with right policies. These policies and actions can reduce greenhouse gas emissions (GHGs) and also energy intensity. By the year 2050, primary energy demand can be reduced by 30 per cent, and carbon emissions by 70 per cent. The main tools for that goal are efficiency enhancement (38 per cent), renewables (32 per cent), Carbon Capture and Storage CCS (7 per cent) and nuclear (12 per cent) according to the IEA report (IEA, 2016a).

After COP 21, new technologies for adoption are required to meet desired emission limits. For developing countries, priority analysis must be done for dependence and also for sustainable economic growth. In this study, an example for priority analysis for developing model is presented. The aim of this study is to use the analytic hierarchy process to do priority analysis as regards to localization of equipment in a power plant. Parameters involved, such as readiness of both infrastructure and human resources, manpower as skilled labour, market potential for equipment developed by transferred technology, and competition in global/internal market, are related to the localization of thermal power plant technologies, and are considered in relation to the country's technological capability, design ability, possession of materials/equipment, and ability to erect a plant.

To develop and maintain a sustainable economy, developing countries need to ensure that the development of both technology and equipment are a priority. It is considered that this study can be used as a starting point for the quick and easy adaptation of foreign technology.

Secure Power Mix Options for Developing Countries

Fossil Power

Coal and gas power plants have a significant portion of total electricity generation (percentage shares of emissions from coal-fired power plants are shown in Figure 5.1), and they are on focus to decrease carbon emissions to meet global temperature rising limit. For the base load, both coal and gas are prior to generate by means of their relative low costs and supplying secure grid. There are some alternative technologies and barriers for coal-fired power generation. Supercritical and ultrasuper critical steam conditions enable more efficiency and lower emission. IGCC power plants are possible where coal characteristics are suitable for this technology. Existing power plants which have low efficiency and relatively small capacity may be stopped when the other sources will be ready. The alternative of coal power is gas fired power plants and nuclear. But nuclear energy is not considered as a clean power generation technology in the scope of COP 21 because of political and social issues (NEA, 2015a). Nuclear energy can be the key source to achieve rising temperature below 2 °C and is the best alternative after coal and gas power after the year of 2030. Renewable technologies grow fast on electricity generation.

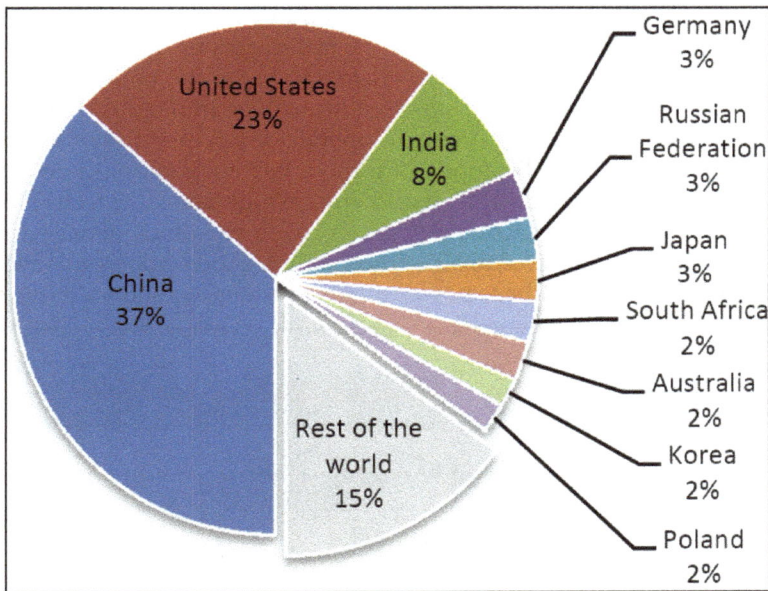

Figure 5.1: Per cent of Global Emissions from Coal-Fired Power Plants (IEA, 2012a; IEA, 2015a).

Coal Power

Coal is the main power generation fuel and has approximately 40 per cent of all generation in the world. It is also currently the largest contributor to emissions in power generation (77 per cent of total CO_2 emissions are from power generation, equal to 10 Gt of carbon dioxide according to the IEA report). In order to keep the global temperature below 2°C in terms of COP 21 a significant decrease in the use of coal without CCS is required especially after the year of 2030 (IEA, 2012a; IEA, 2014a).

Most of coal fired power plants have a boiler with Pulverized Fired (PC) technology. After 1980s, Circulating Fluidised Bed (CFB) technology has become a commercial application not only on low rank coal and also on biomass and waste combustion, with co-firing option. CFB plants are generally considered to have more flexibility than PC ones regarding the composition of the fuel. CFB is a suitable technology for the low rank coal and lignite (spread in India, Turkey, Serbia and Greece; Mills, J. S., 2011) because of high humidity and ash contents (Lockwood, T., 2013). The efficiencies are up to 50 per cent with ultrasuper critical (USUP) boiler technology (currently only 8 per cent in the world) in both PC and CFB power plants. The older boiler technologies are dominant around the world with a 70 per cent share and the most desired option is retirement of these for emission restriction. They are not efficient enough (at a level around 32 per cent) to set emission technologies such de-SOX, de-NOX, disposal particles and CCS (IEA, 2016c).

The coal technologies emit 670 to 880 gCO_2/kWh under good conditions and CCS technologies are required play a vital role in decarbonising the power sector, for the 2DS trajectory. Technological improvements on boiler and emissions to decrease the carbon intensity of coal power plants with new technologies (like USUP and IGCC) could enable coal power plants to operate at lower emissions levels (IEA, 2012a). CCS is one option to be applied for these new technological power plants. After 2030, USUP usage around the world will be high and CCS technologies can be applied on them. Other technological improvements and increased co-firing where sustainable biomass is available (especially a good option for Africa) could help to reduce emissions in these power plants (options are shown in Figure 5.2).

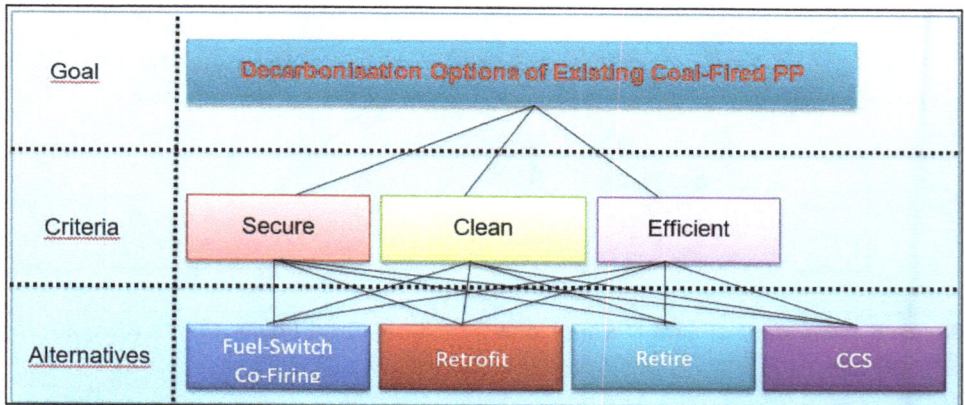

Figure 5.2: Hierarchical Structure of Existing Coal-Fired Power Plants.

Coal is important fuel for China, India, the ASEAN and sub-saharan Africa which are non-OECD countries and also for Turkey which is in OECD (Mills, J. S., 2014). According to IEA reports, governments of these countries should support and prioritise development of CCS needed to meet 2DS targets. The coal-fired power generation without CCS will be the lowest after 2050 and this probably will close at a faster natural infrastructure replacement. Existing coal-fired plant efficiencies should be improved. Retirement (closure of plants) is an option for small and the least efficient units of plants (China is doing this). CCS-retrofitted for high efficient plants and coal-to-biomass conversion (fuel-switch) and/or coal-biomass blending have the technical potential. In existing coal-fired power plants, biomass may be co-fired with coal up to 10–20 per cent for PC and up to 30 per cent for CFB. Oxy-combustion and Chemical Looping Combustion (CLC) are the new technologies for coal-fired power plant to meet clean environment (IEA, 2012a).

Gas Power

Gas power and its operational flexibility plays a critical role in generation since it is balancing daytime fluctuations in power demand and has an important role in the 2DS scenario in the transition to a low-carbon energy system as it can balance fluctuations at increased levels of renewables. This is the best alternative option for nuclear to compete. For the short and medium term, gas fired generation can be the option to reduce emission because of lower carbon intensity comparing to coal power (emitted around half of coal). After 2020 and 2030 respectively, OCGTs and CCGTs will fall and zero-carbon generations such as renewables and nuclear will be the main power mix options (IEA, 2016b).

Gas fired generation requires technological improvements including the application of CCS to be a long time actor up to 2040 in decarbonisation of the power sector. Therefore switching from coal- to gas-fired generation is not enough to limit the temperature to 2°C over the long term and required improvements (IEAGHG, 2012b; IEA, 2016a).

Nuclear Power

To limit rising of the temperature below 2°C requires a mix of technologies including nuclear and renewables in power generation. Nuclear is the best option in short and mid-term by the year 2050 to provide the goal of the 2DS. It is already a mature technology and fourth generation reactors are to have alternatives for heat transfer fluid. Some difficulties with nuclear technology are not technical nor scientific but political and social. Nuclear energy is a low carbon power at 11 per cent share of global electricity and the second largest source after hydro (16 per cent). Nuclear power must be an important component of the world clean energy policy (NEA, 2015a). The world needs a combined strategy of all options for sustainable clean energy including nuclear, renewables, efficiency improvement and energy conservation (NEA, 2015b;IEA, 2016a; IEA, 2016b).

Renewables

Increasing of renewables' share may result in decline of coal-fired power plants. Coal fired-power plant is very important source as a base load. Usage of renewables such as wind, solar and geothermal may result in reliability and supply security. Photovoltaics (PV) panels can be a good source of energy off cities. For the cities, Concentrating Solar Power (CSP) is a good option in sub-Saharan Africa. If the population is distributed in a wide range, PV panels (can be used separately without huge transmission lines) are more suitable than CSP. (IEA, 2016b).

Impacts of New Regulations on Energy Sector

COP 21 and upcoming climate change issues affect energy value chain from primary production to transportation, transmission, storage, distribution and energy demand. Policy related actions refer to the change of all parameters of the energy system to meet climate restrictions and its effects.

The new energy concept covers the following parameters (IEA, 2016b) (related to alternatives as hierarchical structure in Figure 5.3 shows);

* ★ Robustness: ability to withstand extreme conditions,
* ★ Resourcefulness: continuing to effective operations,
* ★ Recovery: comprising to operate at desired performance.

From top to bottom, technological and management measures should be taken in all energy sub-sectors (including energy demand) to adapt to new parameters. Distributed and diversified generation can be solution for reducing the risk coming with increased shares of renewable energy in power mix since the system can have ability to buffer and localise outages. Existing low carbon energy technologies (such as CCS, CSP, geothermal and nuclear) consume more water. The adaptation of these technologies to consume less water can be a solution. PV and wind tend to use less water. Energy efficiency is a key parameter on both generation and demand. In developing countries (notably in Asia), economic growth and urbanisation increases energy demand and require new infrastructures and investments. These new situations require the adaptation and some mitigation actions including improving efficiency and decentralizing renewable energy, deploying technologies such as energy storage, and smart grids in order to supply more flexible, responsive and more climate friendly energy system (IEA, 2016a).

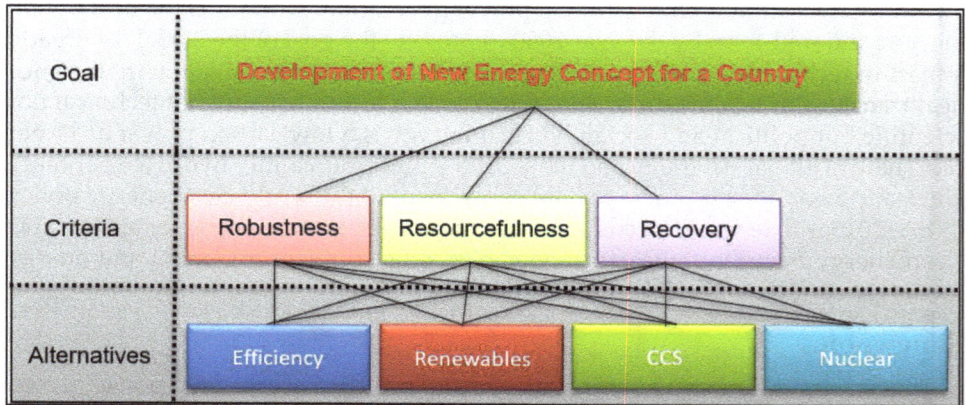

Figure 5.3: Hierarchical Structure of New Concepts.

Opportunities for Developing Countries

Demand is rapidly increasing in many developing countries, where a different set of challenges and opportunities is faced for new efficient technology and low-carbon energy sources. These countries also have important roles to solve the global problem of climate change co-operating with developed countries which are responsible for the existing pollution (Wilkins, 2002).

The new cities in emerging economies are going to play a crucial role for the 2DS goals because of urbanisation. Cities in developing countries can be strategic areas and niches for innovative energy technologies (such as electric vehicles, building-integrated PV panels, heating and cooling) to show from demonstration phase through deployment to commercial maturity by providing the right local and national energy frameworks and policies for local energy sustainability issues. Urbanisation and the growth in primary energy demand (about 90 per cent between 2013 and 2050) will take place in non-OECD countries (NEA, 2015a).

District Energy networks with smarter integrated urban energy grids (micro grid) in these cities can provide a more flexible, cost-effective less carbon-intensive energy because of reducing the need for investments in national energy infrastructure.

Energy needs of urban areas can be met by domestic renewable sources to provide increasing urban energy resilience and retaining economic value. By the year of 2050, 5 per cent of the needs of electricity of urban areas can be supplied by solar PV panels. Because of the lower density of PV panels, they can be suitable in small cities than larger ones. And also PV panels do not require a huge infrastructure investment. Industrial sites can be integrated to energy resources and urban demands. District heating system (taking steam from a power plant or installing a small power plant for heating) can provide flexible, cost-effective and environmental friendly solution in cities of developing countries. As an urgent action, for non-OECD countries, positive CCS approach and related projects can be more efficient than zeroing emission (IEA, 2016a; IEA, 2016b).

Policies for Emerging Economies

Policies should include appropriate strategies and specific actions at all level of energy sectors (such as local and central authorities, investors, operators, producers and researchers *etc.*) especially for the cities and countries to meet climate change. New technologies incorporated in business models and localized for developing countries are required to be developed and deployed. Governments will have to adopt own policies (including financing) to give guidance for private sector action in order to solve their own problems by means of basic science and technology development to support sustainable economic growth (IEA, 2016b).

Developing countries' plans should include solutions that can be actions to reduce carbon emissions and to enhance resilience. This process can identify some opportunities including conventional emission reduction activities, increasing share of distributed renewables, increasing efficiency of both generation and consumption and also developing own technologies suitable for local conditions. Governments have a key role to regulate especially huge investments as public-private partnerships (IEA, 2016b).

Technological Development Models of Power Generation

For adaptation, there are some regional factors including cost of capital, technical feasibilities like levels of solar radiation and wind speed, and suitable plant technologies. Commercial companies around the world sell technologies which are used widely. They need not be the solution for extreme conditions. Developing countries need technologies suitable for domestic sources to generate electricity. They have to develop these technologies or transfer them cooperating with local vendors and research infrastructures (including manpower).

Steps for governments to make policies can be seen in Figure 5.4.

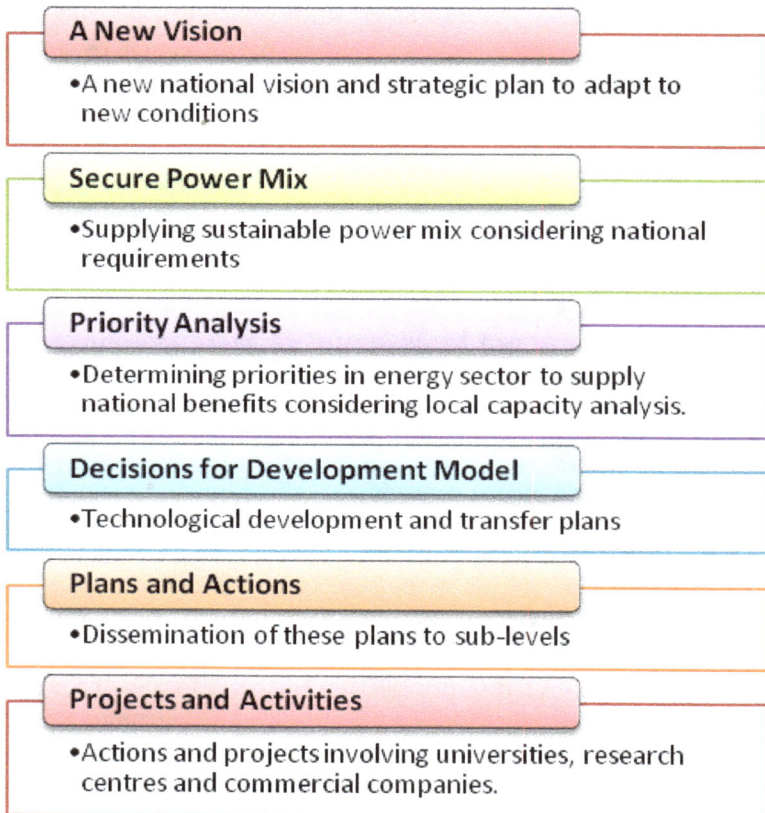

A New Vision

•A new national vision and strategic plan to adapt to new conditions

Secure Power Mix

•Supplying sustainable power mix considering national requirements

Priority Analysis

•Determining priorities in energy sector to supply national benefits considering local capacity analysis.

Decisions for Development Model

•Technological development and transfer plans

Plans and Actions

•Dissemination of these plans to sub-levels

Projects and Activities

•Actions and projects involving universities, research centres and commercial companies.

Figure 5.4: Steps for Adaptation for a Nation.

The Importance of Science and Technology for Developing Countries

Economic growth needs technological development for developing countries in order to produces goods and services in a cost effective manner by competing in the market. Improvements of living standards can be ensured by increased productivity attained through R&D activities. Gross Domestic Product (GDP) measures the impact of science and technology on industry and all other related areas. GDP per employed person also shows availability of natural sources, ability of workforce, level of infrastructure and other social and economic factors (Jain, R.K., 2010).

According to the AHP rating of readiness, manpower, market and competition, technology (42 per cent) is the most important priority for Turkey, followed by Design (26 per cent), Manufacturing (15 per cent), Material/Equipment (11 per cent), and Erection (6 per cent) (as shown in Figure 5.5). It is clear that these results are country dependent and are likely to be different when applied to other countries. Furthermore, the results will also vary between developing countries, as there are many parameters affecting each country's internal factors.

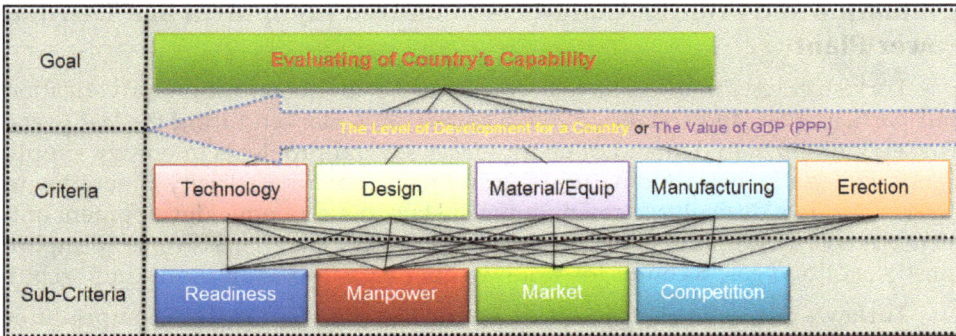

Figure 5.5: Hierarchical Form of Evaluating Capabilities.

Technology Transfer

It is initially much cheaper, easier, and faster to transfer technology from developed countries than to develop; sometimes it is not possible to develop it by developing countries. Such technology can be transferred under suitable contracts. Transferred technologies provide competitive advantages for developing countries. However, transferred technologies need to be adapted according to local conditions, to be absorbed within local companies and deployed for future improvement. The success level of transferring technology depends on the transferee's ability to manage the whole process. The company or country needs to be ready to enhance both its infrastructure and its human skill to be competitive in international markets. The right policies, actions and plans should be adopted by each country by means of priority analysis at all levels because of the time limit (Cohen, 2004; Wilkins, 2002; Yagmur, L., 2016).

Localization of Technology

There is a big gap between developed and non-developed countries. Technology transfer can provide rapid improvements on developing countries to close the gap. But developed technologies may not satisfy local requirements of developing countries. Localization (used as meaning "the act or process of making a product suitable for use in a particular country or region") of a transferred technology has a key role for that country. It has some steps required; adaptation, absorption, development, and deployment of technology. The processes are also affected by the level of technical infrastructure and human resources capabilities. Localization priorities need to consider these factors so that industrialization is accelerated. The system/equipment involved in localization can differ according to the priorities of each country, and success also depends on the situation of certain factors directly related to the country's condition and opportunities related to the nation's development policies and strategies. For developing countries, localization of the electricity generation sector is important to secure the supply of electricity generation and to maintain sustainable economic growth (Yagmur, L., 2016).

Evaluation and Priority Analysis of Localized Equipment in a Thermal Power Plant

Turkish energy markets have been changed with liberalization, privatization and participation, and converted into a competitive market during the last years. Turkey with its high demand of energy faces several problems in energy supply issues. Especially, imported natural gas and hard coal for electricity generation in PPs constitute a difficulty in maintaining sustainable economic development and increases energy supply dependence on developed countries. Turkish energy import, mostly of oil and gas, accounts for around a quarter of its overall annual import bill.

Turkey's domestic energy source among fossil fuels is mainly comprised of lignite reserves. In 2000s, the government has tried to develop a policy to encourage finding of additional domestic lignite reserves by using a suitable model of public-private-partnership in Turkey so as to decrease the rate of imported natural gas for electricity generation. At present the high demand of electricity is mostly met by combined cycle natural gas fired PPs by a share of about 44 per cent in 2013, 48 per cent in 2014 and 35 per cent in 2016 according to the reports. This situation resulted in increasing imported natural gas, mainly from Russia and Iran.

In spite of its poor quality, lignite reserves are spread across the whole country in more than 40 regions. According to the TKI's (state owned Turkish Coal Company) latest data, amount of those resources exceeded 11.8 Gt of lignite. Turkey's lignite has high moisture and ash rate leading to lower calorific value. Most of coal fired thermal PPs in Turkey had been built in 1980s and 1990s before the development of CFB technology. Therefore, most of the existing PPs have Pulverized Coal (PC) combustion technology. The problems in combustion of domestic lignite in PC fired boilers cause decreased availability and capacity factor leading to rise in operation costs (Yagmur, L., 2016).

Turkey has to develop its own combustion system in CFB suitable for domestic lignite by evaluating equipment priorities.

The Importance of Priority Analysis in Emerging Energy Sector

National priorities and dependence can change for each country. It should develop a visionary strategic plan and related projects to meet national requirements. A number of methods and systematic approaches are required to do the analysis, and in this respect AHP is the most powerful and easy method used in decision making, and used to prefer one alternative to another, or to rank all alternatives in view of criteria and sub-criteria, which are known as factors (Yagmur, L., 2016).

Evaluation Methodology

Multi-Criteria Decision Making (MCDM) tools and other related approaches are commonly used to analyse and make right decisions, particularly in relation to issues that include different types of factors. The goal of MCDM is to select the best alternative, or to rank a set of such alternatives. A number of useful and popular MCDM methods exist, such as AHP, TOPSIS, PROMETHEE, and EVAMIX.

Prior Systems/Equipment

In this section, an evaluation and priority analysis is given. The aim is to determine a priority order of equipment used in PPs in relation to localization. This example focuses only on the main components, without consideration of the technological classifications mentioned above. The components are selected in terms of Turkey's localization of technology strategy. The emission control system is not included for evaluation in this study. The eleven items chosen are: Boiler (considering CFB technology), Steam Turbine, Feed Water Pump, Condenser, Circulating Water Pump, LUVO, grinding Mill, feed water Heater), Generator, Fans, and Electric Motors (Yagmur, L., 2016).

There are a number of internal and external factors that affect the success of localizing technology. Five stages can be defined in the process between an initial idea and the final product, and these factors include the technology of a system and the product development process, which consists of the design, material, and manufacturing and assembly (used in the study as "Erection" as widely used in the power sector). Technology, Design, Material/Equipment, Manufacturing, and Erection are given originally as a function of localization for a country in this study. Each factor or stage then needs to be considered in relation to the localization parameters. The technical infrastructure is the most important factor to consider when localizing transferred technology within a company/country. In this study, this factor is known as "Readiness." The second factor is related to human skill or to experienced personnel and is referred to as "Manpower" in this analysis. These internal parameters have vital roles, and as such are the most important issues in relation to the diffusion of technology transfer within a country. However, other external parameters such as "Market" and "Competition" also need to be taken into account (Figure 5.6; Yagmur, L., 2016).

After the analysis and calculations (detailed procedures and calculations are published in; Yagmur, L., 2016). The boiler is seen as the most important part in view of design, manufacturing, and also operation, since its parameters are dependent on local situations and are affected directly by fuel conditions and combustion behaviour. The boiler is also the most important equipment in the priority analysis.

Although the turbine is considered to be important equipment for a thermal PP, the ratings in this study are compiled according to the country's priorities and local conditions. The steam turbine is independent of local conditions and can be readily found globally, and is a standard equipment (mainly depends on steam

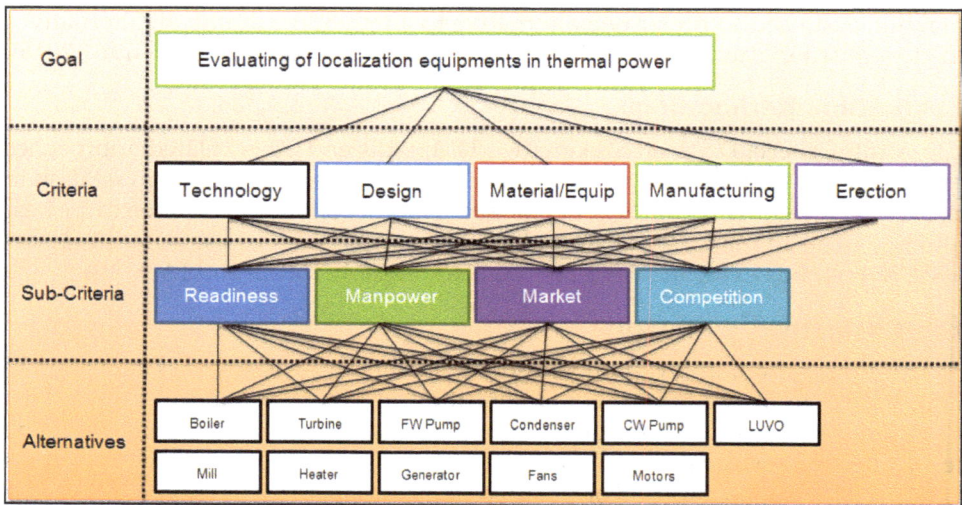

Figure 5.6: Evaluating Structure of Localization Equipments.

pressure and temperature are specified with the steam flow rate) in comparison with the boiler. The other main factors are Market and Technology in relation to the turbine, as Turkey does not yet have turbine technology at the level of a PP, and the market for turbines globally is very active, mature, and competitive. With such considerations, the turbine ranks 8th, as shown in Table 5.1; and heaters, fans and motors are ranked after boiler. This equipment is quite simple compared to other equipment, in relation to the technology involved. The generator and FW pumps are ranked the last. Turkey does not have the technology as yet at a PP generator level, and it is hard to compete for a FW pump at MW power capacity within the market. All other equipment is ranked and listed in Table 5.1 (Yagmur, L., 2016).

Table 5.1: AHP Scores and Priorities

Equipments	Score	AHP Rank	[%]
Boiler	0.202	1	20
Heaters	0.132	2	13
Fans	0.113	3	11
Motors	0.113	4	10
LUVO	0.090	5	9
Condenser	0.084	6	8
CW Pump	0.081	7	8
Turbine	0.078	8	8
Mill	0.052	9	5
Generator	0.046	10	5
FW Pump	0.027	11	3

Conclusions

COP 21 and upcoming restrictions to provide clean environment will result significant changes in energy sector. Strategic plan and actions are required for developing countries because of increasing energy demand and there is need for new technologies. There are many possibilities and technologies to meet climate change limits. Priority analysis is a key action for the rapid economic growth for developing countries. The number of sourcing technologies by developed countries is very high and their market is in developing countries. Governments should take actions considering national and local benefits at different levels. An alignment of international and local activities should be done by means of regulatory frameworks with technological innovations and business models. Sustainable energy policies help governments to implement effective actions. National and local capabilities should be determined in order to decide taking actions especially in technology transfers considering readiness (both infrastructure and manpower). Nationally funded business models and partnerships at different levels will support the economic growth.

In this study, an example for the AHP method is used to conduct a priority analysis using expert judgement, with consideration of the country's local situation, as this approach can be useful for developing countries. Priorities are related to the level and situation of the country's technology, design, material/equipment, manufacturing, and erection, as these are subject to other localization factors such as readiness, manpower, market, and competition.

Abbreviations

2DS:	2°C Scenario
AHP:	Analytic Hierarchy Process
ASEAN:	Association of Southeast Asian Nations
CCGT:	Combined-Cycle Gas Turbine
GDP:	Gross Domestic Product
GPD (PPP):	GDP per Employed Person
CFB:	Circulating Fluidised Bed
GHG:	Greenhouse Gas
CLC:	Chemical Looping Combustion
COP 21:	Conference on Climate Change, agreed in December 2015
CCS:	Carbon Capture and Storage
CO_2:	Corbondioxide
gCO_2:	Grammes of Corbondioxide
CSP:	Concentrated Solar Power
Gt:	Gigatonne
CW:	Circulating Water
FW:	Feed Water
IGCC:	Integrated Gasification Combined Cycle

OCGT: Open Combined Gas Turbine

OECD: Organisation for Economic Co-operation and Development

PP: Power Plant

PC: Pulverized Fired

PV: Photovoltaic

USUP: Ultrasuper Critical

MCDM: Multi-Criteria Decision Making

MW: Megawatt

References

1. Cohen G., 2004, Technology Transfer: Strategic Management in Developing Countries, Sage Publications.

2. IEA, 2012a, CCS Retrofit, OECD/IEA, Paris.

3. IEAGHG, 2012b, CO_2 Capture at Gas Fired Power Plants, OECD/IEA, Paris.

4. IEA, 2014a, Energy, Climate Change and Environment: 2014 Insight, OECD/IEA, Paris.

5. IEA, 2015a, CO_2 Emissions from Fuel Combustion: Highlights, OECD/IEA, Paris.

6. IEA, 2016a, Energy, Climate Change and Environment: 2016 Insight, OECD/IEA, Paris.

7. IEA, 2016b, Energy Technology Perspectives 2016, OECD/IEA, Paris.

8. IEA, 2016c, Technology Collaboration Programmes: Highlights and outcomes, OECD/IEA, Paris.

9. Jain, R.K., 2010, Managing, Research Development, and Inovation, 3. Edition, John Wiley and Sons Inc., New jersey, USA.

10. Lockwood, T., 2013. Techno-economic Analysis of PC versus CFB Combustion Technology

11. Mills, J. S., 2011. Global perspective on the Use of Low Quality Coals, IEA

12. NEA, 2015a, Nuclear Energy: Combating Climate Change, OECD/IEA, Paris.

13. Mills, J. S., 2014. Prospects for Coal and Clean Coal Technologies in Turkey, IEA

14. NEA, 2015a, Nuclear Energy: Combating Climate Change, OECD/IEA, Paris.

15. NEA, 2015b, Technology Roadmap: Nuclear Energy, 2015 Edition, OECD/IEA, Paris.

16. Wilkins, G., 2002, Technology Transfer for Renewable Energy, The Royal Institute of International Affairs.

17. Yagmur, L., 2016. Multi-criteria evaluation and priority analysis for localization equipment in a thermal power plant using the AHP (analytic hierarchy process), Energy 94, 476-482

Chapter 6

Deployment of Biogas Technology for Solving Energy Problems in Nigeria

Nwankwo Nnenna Cynthia

Department of Environmental Sciences and Technology
Federal Ministry of Science and Technology, Abuja, Nigeria
E-mail: nnennasmails@yahoo.com

Abstract

Energy is the essentiality of Nigeria's economic growth and development. As a result, it is pertinent for industrial and domestic utilization. The Biogas generation as a form of energy in the Nigerian energy system is gradually creeping into the country as a power generation source for cooking and heating purposes both in domestic and industrial uses. The Federal Government of Nigeria in her desire to promote the production of biogas has commenced its promotion and development in most part of the country through the Energy Commission of Nigeria (ECN), an agency under the purview of Federal Ministry of Science and Technology (FMST).

Keywords: Energy, Biogas, Production, Development, Nigeria, Biomass.

Introduction

Energy has been and is still the driver of development in different countries of the world. It is considered as the basic factor that contributes to the sustainable development of nations. Energy is also an essential resource for modern industrial societies which is characterized by an intensive consumption of biomass as a source of raw materials with a prediction of an increase by 50 per cent in world energy consumption by 2030 (EIA, 2009). The energy sector is a key important aspect in Nigeria, in addition to its macroeconomic importance, it also has posed to reduce poverty, enhancing productivity and improving the general quality of lives of our people. There are some similarities within the sectors of the economy. On the other hand energy sector contributes to a stable growth in the economy together with the realization of social and political objectives; the modernization and expansion of energy supply system to meet energy demand in future is required for a large amount of human and financial resources.

Over the years, the volume of Municipal Solid Waste (MSW) from various communities has been on the increase daily as a result of rapid growth in population and urbanization. In Nigeria for instance, very few of these wastes are sent to dump sites but the majority ends up in drains, water bodies and open places due to improper waste management systems, leading to drainage channels becoming blocked with solid waste, thus contributing to environmentally related sicknesses like malaria and cholera. Open dumping, open burning, controlled burning and tipping at dumpsites have become popular processes of waste disposal. This has brought about major sanitation problems in towns and cities as they have been inundated with management of municipal solid matter. Untreated and unmanaged biodegradable waste creates odour, and leads to adverse environmental impacts. As a result of this, the utilization of the waste as an important raw material for biogas production is being promoted for the supply of required energy in towns and villages in the country.

Recycling biomass for energy and other uses, would minimize the need for "landfills" to hold garbage. This waste can produce electricity, heat, compost materials or fuels. In Nigeria, Lagos state generates over 8000 tons of waste a year (Solar Energy Society of Nigeria,2003). Also Rivers State in the southern part generates over 5,500 tons of waste per year mainly from food processing, plastics, rubbers, *etc.* Kano generates at least 1,700 tons of waste comes from farming industries while Kaduna produces an estimate of over 3,400 tons of waste from the industries annually. With all these waste been utilized, the 60,800,600 tons of biomass in Nigeria can produce about 2,500MW of electricity. That's enough energy to make electricity for about two million houses.

Biomass, on the other hand, is often offered to gain considerable advantage for limiting and adapting to climate change and to power society indefinitely. Although the transition will take time due to redesign requirement of many aspects of infrastructure systems but this role played by renewables is necessary for each country to be independent from fossil fuels and to respond to the energy shortage crisis. There is now greater emphasis on biogas production and sustainability– a promising mean of generating renewable energy, reducing the organic waste stream as well as achieving multiple environmental benefits.

Figure 6.1 shows there is increase in the population of livestock.

Figure 6.2 shows that there is high production in the cereals and tubers in Nigeria.

Nigeria at a Glance

Nigeria is located on the west coast of Africa and is boarded by The Gulf of Guinea to the South, Niger to the North, Cameroon to the East and Benin Republic

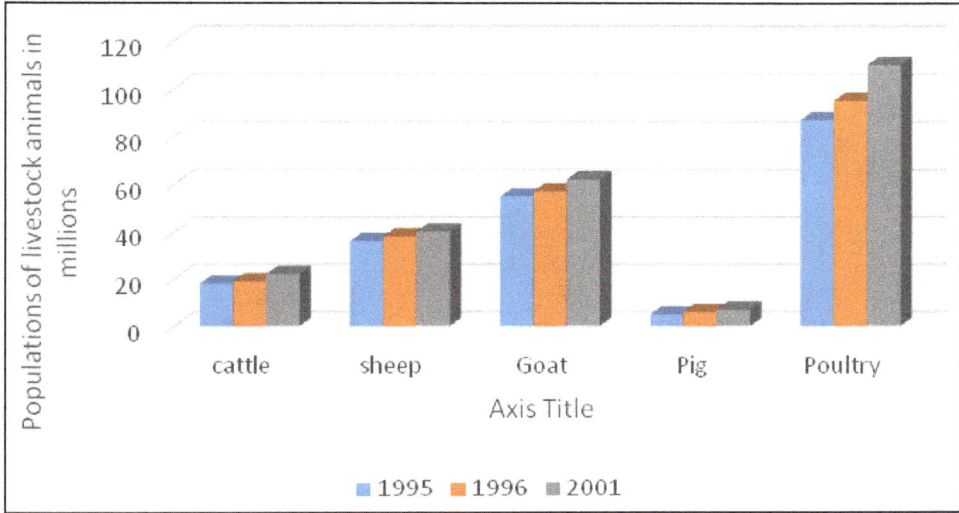

Figure 6.1: Nigeria Livestock Population (*Source*: ECN and UNDP final reports, 2005).

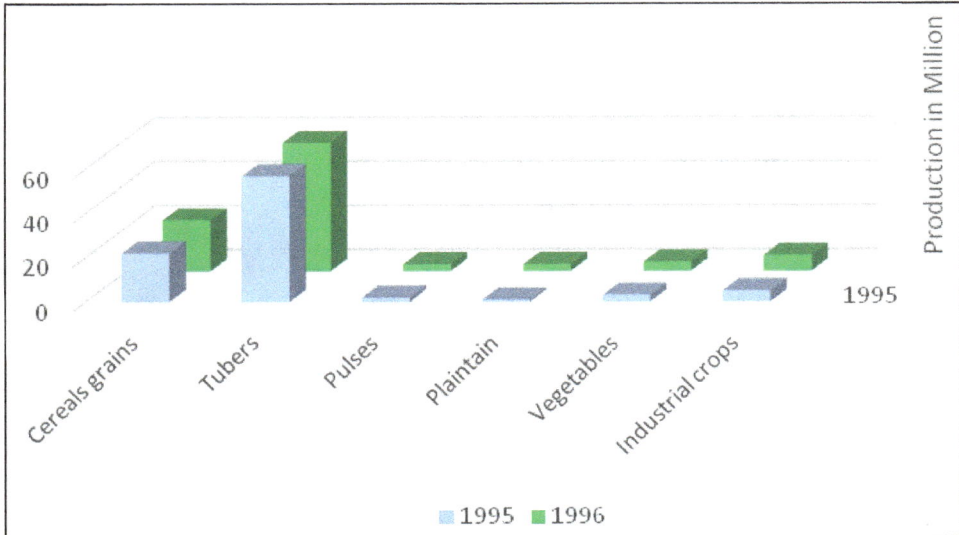

Figure 6.2: Production of Major Crops in Million Tonnes (*Source*: ECN and UNDP final reports, 2005).

to the West. With land area of 923,768 km^2 Nigeria is the continent's most populated country in Africa, with over 180 million people with a growth rate of 2.44 per cent (NPC, 2016). Nigeria lies within latitudes 4.32°N and 14°N and longitudes 2.72°E and 14.64 °E.

Nigeria is along the tropics and it experiences high temperature throughout the year with a mean temperature at about 27°C with an average maximum temperature about 32°C on the coastal part and 41°C in the Northern part.

Figure 6.3. shows the National Grid system in Nigeria with the distribution companies in the 11 states of the federation. The red lines show the transmission grid, and the blue lines show the water flow pattern.

Figure 6.3. National Grid system in Nigeria (*Source*: "Nigerian National Electricity Grid." Global Energy Network Institute).

History of Nigeria Electricity Supply Industry

The Electricity Supply Industry (ESI) in Nigeria dates back to 1866 when two small generating sets were installed to serve the colony of Lagos. In 1951, the Electricity Corporation of Nigeria (ECN) was established through the Act of Parliament to cater for all power supply systems in the country. The Niger Dam Authority (NDA) was subsequently established for the development of hydroelectric power project at Kainji on the River Niger. The two establishments were merged to form the National Electricity Power supply in 1972. The total installed capacity rose from 30MW in 1956 to about 6,000MW in 2006 with a thermal hydro mix of about 70:30.

The per capita consumption of Nigeria ranged from 68 to 95 kWh between 1980 to 1997 which was about 17 per cent of the African average. Furthermore, as in 2006 85 per cent of the 774 Local Government Areas were connected to the national grid, yet only 4 out of 36 had 100 per cent grid electricity, namely FCT, Ekiti, Lagos, and Osun.

Current Energy Consumption

Energy Information Administration estimated that in 2011 Nigeria's primary energy consumption was about 4.3 Quadrillion Btu (111,000 kilotons of oil equivalent) of this, traditional biomass and waste accounted for 83 per cent of total energy consumption. This high per cent represents the use of biomass to meet off-the-grid heating and cooking needs, mainly in rural areas. Nigeria has vast natural gas, coal and renewable energy resources that could be used for domestic electricity generation, yet lacks policies to harness resources and develop new (and improve the current) electricity infrastructure, however the policies in 2012 and 2013 to develop these have addressed this concern.

Energy Policy Issues in Nigeria

Overview of Nigeria Energy Policy: In order to develop and promote a workable energy solution in the country, several policy issues and strategies had been put in place by the government of Nigeria:

Energy Master Plan with a projected increase in energy demand and supply at the optimistic growth rate of 13 per cent;

★ The nation shall effectively harness non-fuelwood biomass energy resources by integrating them with other energy resources.

★ The nation shall promote the use of efficient biomass conversion technologies.

Renewable Energy Master Plan was produced in 2006 with support from the UNDP to promote and integrate non-fuel wood biomass energy resources with other energy resources. The policy is being implemented by Nigeria's Federal Ministry of Environment that aims to increase the contribution of Renewable Energy to account for 10 per cent of Nigerian total energy consumption by 2025. The policy primarily addresses Nigeria's need for increased electricity supply, improved grid reliability and security.

Figure 6.4 focuses on sources of energy types and its reserves, from the Department of petroleum Resources, in which crude oil has 37.2 billion barrels. Solar Radiation of 3.5-7.0 kWh/m²/day shows that Nigeria has high intensity of solar radiation.

Summary of renewable energy targets, Renewable electricity supply projections in mw 7% GDP growth rate

S/N	Resources		Reserves
1	Crude oil		37.2 billion Barrels(DPR,2014)
2	Natural gas		1882.3 trillion SCF(DPR,2014)
3	Coal and Lignite		2.734 billion tonnes
4	Large Hydropower		11,250 MW
5	Small Hydropower(<30 MW)		3,500 MW
6	Solar radiations		3.5-7.0 kWh/m²/day
7	Tar sands		31 billion barrels of oil equivalent
8	Wind		(2-4) m/s at 10m height
9	Biomass	Fuel wood	11 million hectares of land
		Municipal Waste	30 million tonnes/year
		Animal Waste	245 million assorted animals in 2001
		Energy Crops and Agric residue	72 million hectares of agric land
10	Nuclear Element		Not yet quantified

Figure 6.4: Nigeria's Energy Reserves/Capacity as on December 2013.

Summary of renewable energy targets, Renewable electricity supply projections in mw 7% GDP growth rate

S/N	System	Short term	Middle term	Long term
1	Hydro (LHP)	3000	6000	6000
2	Hydro (SHP)	43	533	533
3	Solar PV	1,400	3000	20,000
4	Solar Thermal	-	45	6,000
5	Biomass	5	16	50
6	Wind	20	22	30
	All Renewables (MW)	4,468	10,026	32,613
	All Energy Resources	26,000	52,000	160,000
	% RE	17	19	20

Figure 6.5: Renewable Electricity Supply Projection [*Source*: (i) NNPC (2013): Annual Statistical Bulletin; (ii) TCN (2013): Annual Technical Report; (iii) ECN (2005): Renewable Energy Master plan (REMP): (iv) DPR (2014)].

The renewable energy supply projections with a 7 per cent growth rate for short medium and long term are shown in Figure 6.5. For example, the hydropower in the short term is targeted at about 3100 MW with a 7 per cent growth rate.

National Energy Master Plan (NEMP)

It recommends that;

★ Capacity building and extension of training to local workers on the applications, installation, and maintenance of biomass energy technologies;

★ Providing fiscal incentives to encourage local production of biomass energy systems;

★ Building indigenous capacity in the design, development, installation and maintenance of efficient cook stoves and briquetting machines;

★ Identifying suitable bio-energy based technologies and embarking on intensive R&D activities of the same.

Bioenergy Policy

The energy available is abundant from the bio-energy if meaningfully introduced into the nation's energy mix through the development of a detailed programme. The programme should encompass fully supported research, development, demonstration and manpower training components. Policies recommended are:

★ The nation shall improve measures required to support initiatives aimed at reducing forest thinning and to enhance collection and use of forest residue;

★ The nation shall effectively harness non-fuelwood biomass energy and integrate them with other energy resources;

★ The nation shall promote the use of efficient biomass conversion technologies;

★ The nation shall incorporate waste to wealth strategy in its overall management framework.

Biogas in Africa

Africa is blessed with an abundant of diverse and un-tapped renewable energy resources that are yet to be utilized, for improving the livelihood of the vast majority of the population and developing the economy. The production of biogas via anaerobic digestion of large quantities of agricultural residues, municipal wastes and industrial waste water would be benefitted in the African society by providing a clean fuel in the form of biogas from renewable feed stocks and help abrupt energy poverty. Biogas technology can also serve as a means to overcome energy poverty that poses a constant barrier and threat to economic development in Africa. Anaerobic digestion of the large quantities of municipal, industrial and agricultural solid waste in developing and under developing countries is extremely considerable under perspective of sustainable development. However, the use of biogas is not widespread in Africa at the moment unlike the developed countries of the world. There are many reasons attributed to the economic, technical and non-technical nature for the marginal use of biogas in Africa. The key issue for biogas technology in Africa is to understand why there are some limitations in the

Figure 6.6: A Fixed Dome Bio Digester being Constructed in University of Energy and Natural Resources, Brong Ahafo Region in Ghana.

use of biogas despite demonstration done by the government on the viability and effectiveness of using biogas plants.

Biogas Development and Production in Nigeria

How does Biogas work?

Biogas technology is carried out in a biological engine known as Biogas Plant or Digester. The Biogas plant helps to maintain conditions for natural biological process to take place optimally to produce the desired results. Biogas plants consist mainly of the Bioreactor, Gas storage vessel and the Utility points. Once the process begins it continues indefinitely as long as wastes are added daily in optimum conditions into the biogas plant and as long as integrity is maintained. The process is odourless with the daily production of biogas, fertilizer and mineralized water.

Figure 6.7: Floating Drum Bio Digester in University of Energy and Natural Resourses.

Biogas is comprised mainly of: Methane (54-70 per cent), Carbondioxide (27-45 per cent), Nitrogen (0.5-3 per cent), Hydrogen (1-10 per cent), Carbon monoxide (0.1 per cent), Oxygen (0.1 per cent) and Traces of hydrogen sulphide. Available biomass includes fuel wood, agricultural waste, crop residues, animal and poultry dung, municipal waste (Sambo, 2009). There is an uncertainty in the magnitude of the availability of total biomass in the country. These resources are based on some natural forests and plantation made up from wood volume (Onyegegbu, 2003).

Different Models of Biogas

The biogas plant being produced in Nigeria varies from the domestic type to the industrialized biogas like floating drum bio digester, fixed dome digester and the polyester, but in Nigeria there are no different practices as observed in Ghana. In Nigeria, the Energy Commission has done quite a lot in the areas of this technology. They have centers across the country that promotes the research and development of this technology. They have over the years, developed different models of biogas digesters ranging from the Continuous Chinese-type digester Batch -type digester, Continuous Indian type digester, to Plastic and Metallic digesters, in order to rationalize and popularize construction techniques and enable interested persons possessing minimum technical knowledge to build his/her biogas plant.

A portable bio digester comprises of the inlet, where the feedstock are passed from, the bio digester in an anaerobic process that allows the micro-organisms to act on the feedstock and the gas holder where the quantity of gas is measured and utilized.

Figure 6.8: Portable Biogas Digester at the Sokoto Energy and Research Center in Nigeria.

Figure 6.9a: A Fixed Dome under Construction in the Sokoto Energy Research Center in Nigeria.

Figure 6.9b: Fixed Dome Biogas Plant at Sokoto Energy Research Center.

Figure 6.10: Biogas Plant at the University of Agriculture, Makurdi and a Biogas Burner.

Challenges in the Production and Development of Biogas

★ The major hindrance to the use of biogas technology is high cost of installation.

★ There is need to identify organizations or offices within the state and local government levels that will be entrusted with the sole responsibilities in ensuring the full implementation of these policies.

★ Detailed study on biomass resources is required.

★ Inadequate capacity building of human resource in biomass energy

★ Lack of incentives from the government on biomass energy development. The present administration has a way to address the following challenges in pushing for direct utilization and distribution of biogas.

★ Absence of regulatory framework;

★ Absence of comprehensive RET Policy;

★ High initial cost of Biogas;

★ Inadequate financing schemes for Biogas;

★ Lack of favorable pricing policies;

★ Inadequate public awareness on the benefits of Biogas;

★ Uncoordinated R&D

Conclusions

Nigeria is endowed with appreciable biomass resources. There is now greater emphasis on biogas production and sustainability, a means of generating renewable energy, reducing organic waste stream as well as achieving multiple environmental benefits. Federal Ministry of Science and Technology, Federal Ministry of Environment, Federal Ministry of Agriculture, Federal Ministry of Power, Energy Commission of Nigeria, Nigerian Electricity Regulatory Commission, Department of Petroleum Resources and Standard Organisation of Nigeria, make up some of the institutional framework for the promotion, regulation and standardization of biomass energy and its systems in Nigeria. Efforts are being made to get the energy policy and master plan passed into an energy law.

Acknowledgments

I wish to acknowledge my gratitude to Mr. Anayo Linus, Engr Okechukwu, Akachukwu, Mr. Agoro, the Director Environmental Science and Technology, Federal Ministry of science and Technology Mr. Peter Ekweozoh, Dr. Mrs. Kela and Prof. Dioha Joseph from the Energy Commission Nigeria, for their assistance and encouragement.

References

1. Energy Commission of Nigeria (2012). Renewable Energy Master Plan; Second Edition November 2012.

2. Energy Commission of Nigeria (2013). National Energy Policy; Draft Revised Edition 2013.

3. Energy Commission of Nigeria (2014). National Energy Master Plan; Revised Draft Edition 2014.

4. Hydro, J., and Business, P. (2009). Energy Crisis in Nigeria/: Mshandete, A. M., and Parawira, W. (2009). Biogas technology research in selected sub-Saharan African countries – A review, *8*(2), 116–125.

5. Nwankwo Nnenna. (2015). *Biogas Production for Sustainable Development University of Natural and Energy Resources Ghana 2015 (unpublished M.Sc. thesis).*

6. Ogwo, J. N., Dike, O. C., Mathew, S. O., and Akabuogu, E. U. (2012). Overview of Biomass Energy Production In Nigeria/: Implications and Challenges, *1*(4), 46–51.

7. Olanrewaju, O. O. (2009). Waste to Wealth/: A Case Study of the Ondo State Integrated Wastes Recycling and Treatment Project, Nigeria, *8*(1), 7–16.

8. Oluseyi, O., and Kolawole, O. (2009). Environment Nigeria ' S Energy Challenge and Power Development/: The Way Forward, *20*(3).

9. Presidency, T. H. E., and Commission, E. (2003). Federal Republic of Nigeria National Energy Policy: The Presidency Energy Commission of Nigeria, (April 2003).

Chapter 7

Finding Reasonable Energy for Economic Development towards Green Growth in Vietnam: Policy Challenges in Mobilizing Resources

Pham Quang Tri

The National Institute for Science and Technology Policy and Strategy Studies,
Ministry of Science and Technology, Vietnam
E-mail: pqtri2000@gmail.com

Abstract

In the context of the National Green Growth Strategy approved in September 2012, Vietnam has paid much attention in finding reasonable energy for the nation's economic development. On the way of mobilizing resources for the nation's purposes, many issues have been emerged as policies to stipulate the restructuring and improvement of the economic institution towards more effective use of natural resources, and increasing competitiveness of the economy through enhancing investment in new technology, natural capital and economic instrument, policies to restructure its economy to achieve sustainable development and consider green growth an intrinsic part, while keeping in mind the poor people; recognizing disaster risk caused by utilizing energy as fossil energy, hydroelectricity, nuclear energy, *etc*. Besides, the nation has faced increasingly complex policy challenges that include climate change, disasters and erosion, and industrialization and pollution, apart from rapid population growth, limited foreign investment and public taxation.

Keywords: Natural resources, Resources mobilization, Green growth strategy, Energy efficiency.

Economic Growth in Terms of Resources Mobilization

Economic growth means an increasing of total income or output of an economy (of a country, a region or a sector) during a certain period, usually a year.

In terms of inputs for economic growth, three factors have been concerned: The total investment, employees and Total Factor Productivity (TFP), which is essentially: technological advances, techniques, effective use of inputs as labor and capital. The growth based on the first two factors is the width growth (towards volume increasing), and that based on the third factor is called depth growth (towards quality increasing). Thus, economic growth is considered under two angles: Quantity and quality.

In terms of quality of growth, main criteria were evaluated primarily through: the transformation of the scale, structure, quality of the economy in association with guarantee for progression, social justice, rational use of resources and environmental protection.

The quality of economic growth is reflected in the following aspects: (1) a high growth rate maintained for a long period, (2) high productivity, high capital efficiency and enhanced competitiveness of the economy, (3) progression in economic structure shifting; (4) Growth with the settlement of social issues, progression, social justice and ecological environment protection.

Thus, how to mobilize resources for a sustainable growth in line with not only quantity development but also for a better quality growth?

Resources and Resources Mobilization

Resources, in the literature, have been variously defined over time. Steen (2010) has cited the term defined by Penrose (1959) to describe categories of physical and human resources, by means of "..that are 'inputs' in the production process". Meanwhile, Barney (1991); Conner (1991); Wernerfelt (1984) have also indicated that resources are theorized as immutable inputs in the production process. Despite these definitions of "resources as building blocks, surprisingly little progress has been made in the past quarter of a century in defining what these things are" (Steen, 2010). Resources are not only defined as to be solid, durable and independent from the production process, but also they are considered by means of invisible things such as commitment of society or willingness or power of a community in implementing a common plan.

Resources are widely agreed (Bourrelier *et al.*, 1992) as generally being categorized as tangible, intangible, or coalition building. In their argument, Bourrelier *et al.* (1992) defined "Tangible resources are the material resources needed by all activist groups in furthering their causes such as money, space, and a means to publicize the existence of the group and its ideas, while intangible resources include information–dissemination tools such as action alerts, and opportunities to volunteer".

In the field of energy, two accepted sources of energy are renewable energy sources and non-renewable stocks. Mineral fuels (coal, lignite and peat, hydrocarbons, uranium) are the main energy sources occurred in the form of deposits and reservoirs of varying size, namely non-renewable stocks. People have tried to assess these stocks and they have achieved somehow result in recent years. With this result of assessment, they came into a recommendation to save the exploitation of non-renewable sources of energy and replaced by renewable sources of energy. Renewable sources of energy are categorized in wind, solar radiation, hydropower and biomass based energy such as bio gas, bio diesel and heat-power generated from waste. In this situation, the way by which renewable sources of energy are exploited in replacement for non-renewable sources of energy is called mobilization of renewable sources of energy.

Generally, resources mobilization is the way and the means of making resources available, including activities of R&D operations, exploration efforts, capital expenditure for deliveries to consumer markets. Philosophically, a term that is widely used to describe activities that are employed to facilitate movement is mobilization (Stewart-Amidei and Kunkel, 2000). In developing these arguments, facilitating movement is not only focusing on resources term itself, but also impact on behavior of involved partners. Therefore, in the situation of research on energy, resources mobilization also means incentives and regulations for involved partners including supplier side and consumer side of energy to behave in expected way. For suppliers, incentives and regulations should direct them to use renewable sources of energy, while for consumers policies should focus on saving energy and using energy efficiently.

The process of resources mobilization, in the situation of research on the application of green technologies, would be more clear if we focus on the following 6 aspects:

1. Political aspect: The mobilization should clear the issues of: Do people recognise the problems? Do leaders recognise the problem? Which party will support us? Which party will not support us ?

2. Economic aspect: The mobilization should clear the issues of: How is the budget? Which sources? How much? How long? How we are affected by economic trends? What is the economic effect of applying green technologies? Is competition better or worse ? How are our customers affected by economic factors? Is the price of products increased from the application ? Do we solve difficulties related to cost-benefit, risk? How much is pay-back period ?

3. Sociological aspect: The mobilization should clear the issues of: Has culture impacted on behaviour of people or not? How are we affected by social trends? Who would participate in the process? Are we affected by educational trends? Is there enough diversity?

4. Technological aspects: The mobilization should clear the issues of: What sort of technological trends affect the application (by means of technical, systems, processes, software)? What are the prior technologies? For suppliers or for consumers? How do we utilize technology? What could

be done better? How do we monitor? Which class of technologies? Which fields?

5. Legal aspect: The mobilization should clear the issues of: What legal implications can affect our work (might be issues of health, safety, compliance, training, financial...)? What external legal changes can affect us? What are the international regulations?

6. Environmental aspect: The mobilization should clear the issues of: What are the environmental advantages (positives/negative) from applying green technologies? How can we measure/evaluate the environmental impact? How can we be more productive? How can we cut down on waste?

The answer of questions in the six aspects above would provide a general picture of solutions to mobilize resources for enhancing the application of green technologies in a specific circumstance.

Economic Growth of Vietnam

In the last 30 years (1986-2016), the uninterrupted progression of Vietnam's economy has been observed with high-speed, impressive growth. During 1991-2000, the growth rate of the economy was 7.4 per cent/year and in the next 10-year period from 2001 to 2010 it was 7.2 per cent/year.

The impressive features of the economic growth are:

1. The size of the economy has grown significantly. The Gross Domestic Product (GDP) in 2010 was two times higher compared to that of 2000, but at current prices 3.4 times higher. Export turn-over increased more than 4 times/year in 2000. In 2010, GDP/person/year reached 1,200 $ US (in 1990 it was 100 $ US).

2. Economic growth has brought Vietnam into group of developing countries with low average income according to the World Bank's ranking.

3. Solving well the pressing social issues: labor, employment, income, improved life quality, reduced poverty, social security, progress and social justice performed gradually, *etc.*

4. Step by step, Vietnam's economy has integrated deeply and fully into regional economy and the global economy.

However, economic growth in recent years has been mainly in the quantity terms, mainly based on the exploitation of natural resources, increased capital investment and cheap labor. According to the Ministry of Planning and Investment the industry, the growth of which based purely on exploitation and use of natural resources such as agriculture, forestry, fisheries and mining, occupied a large proportion of GDP (about 30 per cent) in the period 1991-2009. Processing and assembling industries which have high intermediate costs, as dependent on imports spending foreign currency, often pollute the environment and not got apparent growth in the service sector. Export turnover has increased rapidly, but the export structure has changed slowly, mainly raw goods, processing product, and mineral resources; so the competitiveness of export is low.

Economy Reform towards Sustainable Development

Vietnam has launched the strategic orientation of sustainable development, with the basic goal of ensuring the harmony between man and nature; development must have three aspects: economic - social - environmental protection. It is interpreted as follows:

1. To ensure harmony between speed and quality of growth.
2. To promote the development of processing industry, manufacturing and supporting industries.
3. To promote rural industrialization, and vigorously develop service industries with high added value and improve techno-economic infrastructure.
4. To develop and expand the domestic market, as well expanding the export markets.
5. To formulate a close link among economic growth and progression, social justice and environmental protection.

Vietnam's National Green Growth Strategy

For the objectives of ensuring fast, efficient and sustainable growth while making a great contribution to poverty reduction and improving the well-being of all people, leading to increased investment in conservation, development and efficient use of natural capital, reduction of greenhouse gas emissions and improvement of environmental quality, and thereby stimulating economic growth, the Prime Minister of Vietnam has approved the National Green Growth Strategy on 25th September 2012.

One element of that strategy is related to improving energy performance and efficiency, reducing energy consumption in production, transportation and trade.

1. Innovate technologies; apply advanced management and operation procedures for efficient and effective use of energy in production, transmission and consumption, especially in high capacity and energy consuming manufacturers.
2. Improving energy efficiency, reducing fuel consumption in transport through technological innovation, regular maintenance of machinery and transport equipment, disseminating eco-driving skills.
3. Establish and publicize standards on fuel consumption rate, roadmap to remove obsolete and energy consuming technologies in energy sector (production and consumption system).
4. Develop a legal basis preparing for the application of technologies to capture, restore and trade various types of greenhouse gases.

Mission for policy makers, researchers, economic entities, managers,… is to find reasonable tasks to fulfill above requirements. Thus, resources mobilization for energy efficiency would be one of those tasks. Through serious research on theoretical documentations and practical experiments, the following five fields have been concerned as focal areas for EE in Vietnam.

1. Energy and Public Policy
2. Energy efficiency in Cities and Buildings
3. Transportation - Cities – Management
4. Energy Efficiency in Industries
5. Renewable Energy and CO_2 emission quota

Resources Mobilization towards Energy Efficiency and Energy Saving

For Mental Sources

Awareness of Society in EE

On papers, Vietnam has paid attention on issuing regulations for energy use in building (Decree 2003 on energy use and saving), concerned to:

1. Person in responsibility for managing the energy use of the building
2. Monthly announcement on energy consumption
3. Issuing action plan for energy efficiency
4. Investing for energy efficiency and monitoring

In addition, regulation for energy used in industry has also concerned of:

1. Technologies for energy efficiency
2. Including Energy Plan in Industrial Planning Concerning the policies for energy use in urban areas
3. Functional activities
4. Daily life energy consumption Policy for energy use in rural areas
5. Focusing on Energy Efficiency
6. Exploring Renewable Sources for Energy: small scale technologies
7. System of indicators

However, the efficiency of the above is still far from expectation.

Political Willingness of Leaders

Recognising the importance of implementing EE, Vietnam government has also committed strongly to develop awareness by conducting National Target Programs, in which initiatives for application of green technologies have been pointed out.

Pressure of International Community

Pressure of International Community has also been a different way to mobilize donor resources for conducting EE. These focus on implementing International Standardization, or "Achievement in managing green techs" or "Success samples in EE"are also good practice for learning. In addition, participating in the global game, Vietnam should accept international regulations on setting rules for CO_2 emission reduction quotas or specific regulation for some products.

Effort of Potential Users

One of important invisible indicator for mobilizing resources in applying green technologies is the availability of initial applicators. This indicator is reflected by their commitment in EE or their participation in industries and service (hotel and transportation).

For Physical Sources

Different kinds of Renewable Energy Sources

The following renewable sources of energy have been recognized as replacement for petroleum energy: Wind, Solar radiation, hydropower, biomass based energy such as bio gas, bio diesel and heat-power generation from waste. The mobilization of these sources, is also a chance and challenge for green technology. The applying process contains not only technical issues but also many other concerns such as:

Financial Supports

Mobilization of financial resources is concerned as the most important issue. Currently, the mobilization does not only depend on state budget, but also call for participation of non-state entities. In some specific project, a Public-Private Joining Cooperation or International support would be a favorable solution.

Human Resources

In order to conduct solutions, human resource is inevitable factor.

Technical Support

Technological application is the main issue in the set of solutions for mobilization of resources for implementing EE in the country. In this way, it does not only focus on technological solutions, but also to popularise the good practices concerning energy efficiency, or accepted technologies.

Education and Training

In mobilizing all resources in the society, in the long term, education and training should focus on enhancing public awareness on energy efficiency by mass media, the communication tool.

Critical Miletones in Implementing Energy Efficiency

National Target Program on Energy Saving and Efficiency in 2006

Vietnam has conducted its national target program on energy saving and efficiency from 2006. The program aims mainly at:

1. Enhancing the role of government in management, decentralized management system
2. Education and communication through mass media for students, applying sample models of energy efficiency in household
3. Development and dissemination of high-performance energy saving equipment, application norms and labels on the products to gain energy efficiency, as well as technical support for manufacturers
4. Energy efficiency in enterprise, with sample model of technical management
5. Using energy in buildings
6. Optimising public transportation for energy efficiency

10 Initiatives on Clean Energy

In an effort to provide a view on the process of mobilization of resources to implement EE in Vietnam, some national level initiatives are presented below:

GE Fixed Series Compensation Solutions

GE is helping economic growth in Vietnam with a brand new custom designed fixed series capacitor banks. The new capacitor banks will help VN deliver more energy without costly upgrades all along our power supply chain, to improve quality and safety while reducing the amount of energy wasted on power lines.

Energy Efficient Public Lighting in Cities (July 2011)

There is a $15 million project plan funded by the UNDP, the Global Environment Facility, local and central governments, and the private sector to bring energy efficient lighting to schools, hospitals, and streets throughout Vietnam.Lighting accounts for approximately 25 per cent of all electricity consumed in the country.

Financing Program for Energy Efficiency and Cleaner Production

This project is a major part of the International Finance Corporation's (a World Bank Group) Global Sustainable Finance Program. Working with a number of selected banks to build sustainable energy portfolios it ooks to promote much greater renewable energy, energy efficient, and cleaner production methods as well as awareness, and to reduce carbon dioxide emissions as well as improve the utilization of natural resources by using available financing for sustainable energy investments. It targets enterprises that are looking to upgrade inefficient production systems and introduce new and clean technologies that will help them reduce their costs and raise their productivity and environmental performance through increased energy efficiencies.

Vietnam Energy and Environment Partnership (2009-2012)

It is a primary component under the brand new support program on climate change adaptation and mitigation. It aims at improving energy efficiency in Vietnamese enterprises to contribute further to a low carbon economy and sustainable development.

Wind Energy Investments in Binh Thuan

16 wind energy projects with investments from 12 foreign and domestic businesses to generate a total capacity of over 2,000 MW have been granted.

A 25 wind-turbine project created by the Vietnam Renewable Energy Joint Stock Company was just completed contributing almost 50 million kWh to the national electricity network.

Climate Projects in Vietnam

The World Bank has been working on a number of climate projects throughout Vietnam to help Vietnam become more energy independent and spend less money increasing the supply of electricity to Vietnam's national grid from renewable sources of energy on environmentally, socially, and commercially sustainable bases.

Building of a 40 Megawatt Biomass Plant (Oct 2010)

Vietnam High Tech Renewable Biomass Energy Joint Stock Company will construct the plant in Phong Chau in the Phu Ninh district. Biomass is seen as a viable energy source in Vietnam as the country is a leading producer of rice and has numerous resources, including rice husk.

Destination for Clean Energy Investors

US First Solar Group announced its plans to build in Ho Chi Minh City a solar panel factory which will creat around 600 new jobs in the country at a total investment capital of around $1 billion. IC Energy has set up a solar panel factory in Chu Lai Open Economic Zone with an investment of over $390 million. German company Roth and Rau invested $275 billion for a factory in Hoa Lac High Tech Park.

Vietnam Renewable Energy Development project

In May of 2011, the State Secretariat for Economic Affairs in Switzerland provided non-refundable assistance to the tune of $2.43 million. It aims at development of clean energy and increasing the supply of electricity from renewable energy sources to the national grid in Vietnam.

International Renewable Energy Exhibition

In March 2012, Hanoi hosted the second international exhibition for renewable energy and decentralized energy solutions to strengthen cooperation, increase understanding of advanced technological renewable energy solutions, and attract investment. It focused on wind, solar, geothermal, hydropower, and bioenergy solutions. Foreign experts and scientists discussed major issues, including current situations and trends, the renewable energy market, and various renewable energy technologies there.

Conclusions

Energy efficiency is only one way for people to show concern about greener growth. Besides, reducing risk from wrong use of energy as hydropower causing heavily flooding for instance, or even risk of atom power causing total disaster, and so on, should be considered to manage human-made risks for our world. A reasonable strategy for mobilizing resources for taking advantages of that energies and reducing risks might be the concern for the whole community.

References

1. Amidei, C. (2012). "Mobilization in critical care: A concept analysis." Intensive and Critical Care Nursing.

2. Bourguinat, É., N. Đ. Hiệp, et al. (2009). Conference Paper. Chính sách sử dụng hiệu quả và tiết kiệm năng lưởng ở Việt Nam É. Bourguinat. Ho Chi Minh City.

3. Bourrelier, P.-H., X. B. d. l. Tour, *et al.* (1992). "Energy in the long term. Mobilization or laissez-faire." Energy Policy.

4. Chu Van Cap (2016). "Reform of Vietnam's economic growth model following to viewpoint of Communist Party Congress IX". http://www.dangcongsan.vn/tu-lieu-van-kien/tu-lieu-ve-dang/sach-chinh-tri/books-0105201511342446/index-210520151130204632.html

5. Pattie, C. and R. Johnston (2012). "Personal mobilization, civic norms and political participation." Geoforum.

6. Steen, J. (2010). "Actor-network theory and the dilemma of the resource concept in strategic management." *Scandinavian Journal of Management.*

7. Villanueva, J., A. H. V. d. Ven, *et al.* (2012). "Resource mobilization in entrepreneurial firms." *Journal of Business Venturing* 27: 11.

Chapter 8

Urban Transport Policies and Programmes in India to Tackle Climate Change

Meenu Galyan

Centre for Science and Technology of the Non-aligned and
Other Developing Countries (NAM S&T Centre)
New Delhi, India
E-mail: meenu4sweet@gmail.com, meenugalyan07@gmail.com

Abstract

The 21st century is witnessing rapid urbanisation at unprecedented rates and amidst increased environmental vulnerabilities to climate change. Increasing urbanization has led to rapid growth of large cities, which are crippled by the sudden rise in travel demand. Transportation sector emissions are one of the major drivers of climate change which accounts for 23 per cent of global CO_2 emissions. This paper gives an overview of the relevant policy initiatives and programmes related to the urban transport sector that have been effective in tackling climate change based on the literature.

Keywords: Urbanization, Climate change, Emission, Transport sector, Policy.

INTRODUCTION

The Intergovernmental Panel on Climate Change (IPCC) in its fourth assessment report concludes that it is extremely likely that the rise in global atmospheric temperature that has taken place since the mid-nineteenth century has been caused by human activities (IPCC, 2007). The IPCC predicted an average temperature rise of 1.5–5.8 °C across the globe during the 21st century, accompanied by increased extreme and anomalous weather events including heat-waves, floods and droughts (IPCC, 2001).

Climate change is a global challenge and different sectors (transport sector, building sector, electricity sector, *etc.*) are significant sources that contribute towards increasing GHG emissions. Two sectors produced nearly two-thirds of global CO_2 emissions in 2013: electricity and heat generation, by far the largest, accounted for 42 per cent, while transport accounted for 23 per cent (IEA 2015).

The 21st century is witnessing rapid urbanisation at unprecedented rates and amidst increased environmental vulnerabilities to climate change (Mittra, S., 2016). Climate change and urbanisation are bound together in a circular positioning of cause-and effect. Urban centres are vulnerable to climate change impacts; and these impacts, in turn, are aggravated by increased urbanisation and its associated carbon emissions. (Mittra S., 2016) By most indications, India's future seems to be urban (Seto *et al.*, 2012). According to reliable estimates, the country's urban population will increase by half a billion over the next four decades (or nearly one million a month) (Economist Report, 2015). As per the UN estimates, India's urban population will reach 600 million by 2030 (40 per cent of the total population) (CASI Report).

India has in fact pursued several initiatives for clean energy development and certain adaptation policies to build resilience against climate adversities. (Mittra S., 2016) For instance, the National Action Plan on Climate Change (NAPCC, 2008) sets out eight missions that align with the goals of combating climate change and pursuing sustainable development. In the recently concluded climate negotiations in Paris, India attempted to manoeuvre the bargaining between economic development and climate action for a win-win situation [Trivedi, 2015]. Recent government initiatives to address urban development and climate change include Atal Mission for Rejuvenation and Urban Transformation (AMRUT) and Housing for all by 2022. (Mittra S., 2016).

Urbanization

Urbanization is a population shift from rural to urban areas, "the gradual increase in the proportion of people living in urban areas", and the ways in which each society adapts to the change [Mittra, S., 2016]. Urbanization has been extremely rapid in the past century. About 75 per cent of people in the industrialized world and 40 per cent in the developing world now live in urban areas (IPCC, 2007). A parallel trend has been the decentralization of cities – they have spread out faster than they have grown in population, with rapid growth in suburban areas and the rise of 'edge cities' in the outer suburbs (IPCC, 2007).

According to the 2011 Census, urbanization in India began to accelerate faster than expected. Since independence, the absolute increase in the urban population was higher than that in the rural population. Population residing in urban areas in India, according to 1901 census, was 11.4 per cent (IPCC, 2007). This count increased to 28.53 per cent according to 2001 census, and crossing 30 per cent as per 2011 census, standing at 31.16 per cent. According to a survey by UN State of the World Population report in 2007, by 2030, 40.76 per cent of country's population is expected to reside in urban areas. (IPCC, 2001) As per World Bank, India, along with China, Indonesia, Nigeria, and the United States, will lead the world's urban population surge by 2050. (Singh, K.N., 2012)

India's Present Urban Challenges

India's present urban challenges make the cities more vulnerable to losses that might result from the impacts of climate change (Divya Sharma and Sanjay Tomar, 2010). Climate change risk to Indian urban centers can be seen in the perspective of the expected changeover in city growth (Divya Sharma and Sanjay Tomar, 2010). By the 2060s, it is expected that there will be approximately 500 million additional people in an estimated 7,000–12,000 urban settlements, with related environmental transitions in water, sanitation and environmental health, air and water pollution and climate change. (Mc Granahan *et al.,* 2007) The cities in India are already struggling with inadequate provision for water, sewerage systems, drainage and solid waste management facilities. Numerous cities lack proper road infrastructure and efficient public transport facilities.

Tabel 8.1: Total Number of Registered Vehicles in India

As on 31st March	Two Wheelers	Cars, Jeeps and Taxis	Buses	Goods Vehicles	Other Vehicles	Total Vehicles
	(as per cent of total vehicle population)					*(Million)*
1951	8.8	52	11.1	26.8	1.3	0.3
1961	13.2	46.6	8.6	25.3	6.3	0.7
1971	30.9	36.6	5	18.4	9.1	1.9
1981	48.6	21.5	3	10.3	16.6	5.4
1991	66.4	13.8	1.5	6.3	11.9	21.4
2001	70.1	12.8	1.2	5.4	10.5	55
2002	70.6	12.9	1.1	5	10.4	58.9
2003	70.9	12.8	1.1	5.2	10	67
2004	71.4	13	1.1	5.2	9.4	72.7
2005	72.1	12.7	1.1	4.9	9.1	81.5
2006	72.2	12.9	1.1	4.9	8.8	89.6
2007	71.5	13.1	1.4	5.3	8.7	96.7
2008	71.5	13.2	1.4	5.3	8.6	105.3
2009	71.7	13.3	1.3	5.3	8.4	115
2010	71.7	13.5	1.2	5	8.6	127.7
2011	71.8	13.6	1.1	5	8.5	141.8
2012	72.4	13.5	1	4.8	8.3	159.5

Source: Offices of State Transport Commissioners/UT Administrations.

Note: 'Other vehicles' include tractors, trailers, three wheelers (passenger vehicles)/LMV and other miscellaneous vehicles which are not classified separately

Urban Transport Scenario in India

Overall population growth and increasing urbanization have led to rapid growth of large cities, which are crippled by the sudden rise in travel demand. The supply of transport infrastructure and services, by comparison, has lagged far behind demand. (Puchera, J, *et al.*, 2005) Urbanisation has also led to enhancement of motorisation. In fact, the growth rate in the number of vehicles has been much faster than that in the population itself. Table 8.1 shows the growth in the number of registered motor vehicles from 1951 to 2012.

Growing per capita income levels and inadequate public transport system have contributed to the increase in demand of personal vehicles. (Verma, A., *et al.*, 2015) India's most acute urban transport problems are not because of the number of vehicles but the high concentration of private vehicles in a few selected cities (NTDPC, 2013). About 32 per cent of motor vehicles are in metropolitan cities alone, which constitute just around 11 per cent of the total population (MoRT and H, 2012).

Transport Sector in India – Contriution of GHG Emissions

Transport sector is the second largest consumer of energy in India. Transport Sector in India is a very extensive system comprising different modes of transport like roads, railways, aviation, inland waterways and shipping, which facilitates easy and efficient conveyance of goods and people across the country (MoRT and H, 2012). According to Eleventh Plan (Planning Commission), railways and roads are the dominant means of transport carrying more than 95 per cent of total traffic generated in the country. Although other modes of transport such as coastal shipping and inland water transport are expected to play an important role in future, the railways and roads would still continue to dominate the transport landscape of the country. The transport sector, overall, is responsible for about 10 per cent of the total energy demand, especially dominating the growing demand for oil in India.

According to International Energy Agency (IEA) 2013, India's transportation energy use would grow at the fastest rate in the world, averaging 5.1 per cent per year, compared with the world average of 1.1 per cent per year. The transport sector comprising of road transport, aviation, navigation and railways accounted for 142.04 million tons of CO_2 eqv emissions, *i.e.*, 7.5 per cent of the total GHG emissions in the country in the year 2007. Of this, road transport alone accounted for 87 per cent of the GHG emissions (*i.e.*, 123.57 million tons of CO_2 eqv) (Ramanathan V. *et al.*, 2014). The transport sector represented 8 per cent of energy in 1990 and will reach 14 per cent in 2035 under NPS, a small but significant growth, as 90 per cent of transport energy consumption will be based on oil (IEA, 2012).

The vehicle-wise share in overall energy consumption in 2010 in the road transport sector is shown in Figure 8.1. Under Business as usual (BAU) scenario CO_2 emission from road transport in India will increase from 19.80 to 93.25 million metric tons of carbon equivalent in 202-2021(Singh, 2006). In 2007, India consumed 595 Mt of energy, and energy-related CO_2 emissions reached 1324 Mt, ranking India the 5th major GHG emitter in the world (MoEF, 2010).

Fuel consumed

- Buses (CNG) 1%
- Buses (Diesel) 13%
- 2w-4s (Gasoline) 16%
- LMV (Gasoline) 4%
- LMV (CNG) 2%
- LMV (Diesel) 8%
- Car+Jeep+Taxi (Diesel) 6%
- Car+Jeep+Taxi (Gasoline) 7%
- Car+Jeep+Taxi (CNG) 0%
- 2w-2s (Gasoline) 2%
- Truck (Diesel) 41%

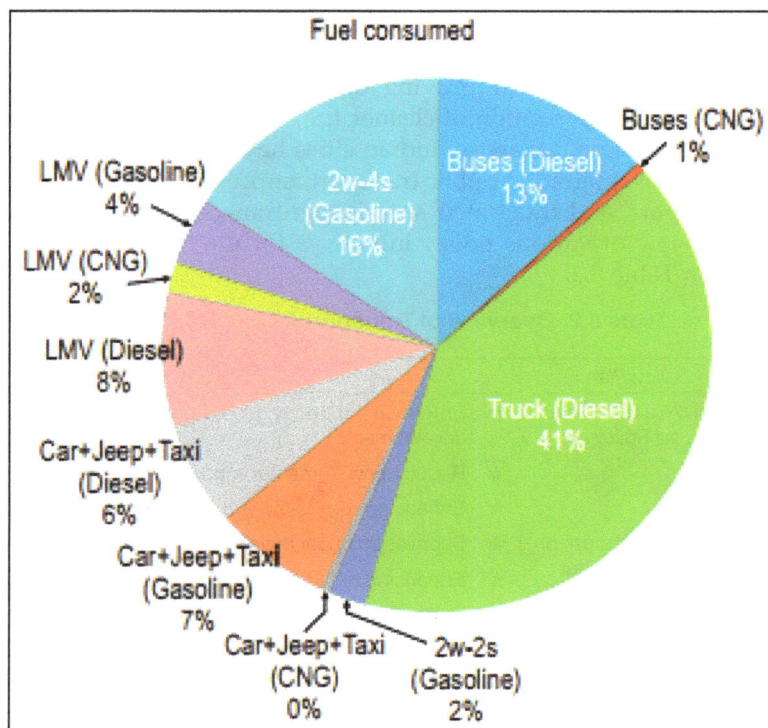

Figure 8.1: Vehicle-wise Share in Overall Energy Consumption in the Road Transport Sector in India (2010) (*Source*: TERI estimates validated with overall consumption in MoPNG, 2013).

Transport sector as a source of GHG emissions not only drives climate change but also contributes to air pollutants in atmosphere and these pollutants have significant adverse human health effects. However, in the urban areas most of the human health impact assessments are based on the levels of PM^{10} in the atmosphere.

Historical and Present Persepctive: Plans and Policies in India (Dhar S *et al.*, 2015)

Productive responses to climate change require policies that consider mitigation and adaptation central to the transportation planning and integrate climate change in the overall policy framework. National transport policies are developed based on the variety of transport demand, the range of fuel supply options, the mix of modes and vehicle technologies, and corresponding infrastructures. The transport system attributes vary at the national and sub national levels, and so do the policy interventions. Transport decisions interface with many other development policy domains, *e.g.* land-use, energy, environment, technologies and finance. The transport decisions have inherent long-term lock-ins enduring several decades.

The transport policy landscape in India has developed extensively with the implementation of several national and sub national policies with the objective of enhancing passenger mobility, improving logistics of freight transport, improving efficiency, promoting penetration of cleaner fuels and vehicles, and reducing air pollution and congestion. Likewise, Indian cities have successfully implemented mass transit systems, upgraded public transport, improved infrastructure for non-motorized transport, and integrated sustainable transport measures into urban plans. An overview of various policy initiatives for the transportation sector in India is given in Table 8.2.

Table 8.2: Overview of Transport Policies in India

Sector	Policy/Plan	Highlights
Urban Transport	National Urban Transport Policy	★ Enhancing mobility to support economic growth and development
		★ Reduce environmental impacts
		★ Enhancing regulatory and enforcement mechanisms
	National Mission on Sustainable Habitat	★ Submission under India's National Plan on Climate Change
		★ Enhancing public transport is one of the key focus areas
	Smart City and AMRUT Programs	★ To develop 100 smart cities
		★ Rejuvenating and revitalizing 500 cities
Alternate Fuels and Vehicles	National Policy on Biofuels	★ Proposed blending target of 20 per cent blending of biofuels, both for bio-diesel and bio-ethanol by 2017
		★ Financial incentives
		★ Waiver on excise duty for bio-ethanol and Excise duty concessions for biodiesel
	National Electric Mobility Mission Plan	★ Investments in R and D, power and electric vehicle infrastructure
		★ Savings from the decrease in liquid fossil fuel consumption
		★ Substantial lowering of vehicular emissions and decrease in CO_2 emissions by 1.3 per cent -1.5 per cent compared to BAU in 2020
		★ Phase-wise strategy for Research and Development, demand and supply incentives, manufacturing and infrastructure upgrade
Intercity Passenger Transport	High Speed Rail Project	★ To develop High Speed Rail corridors in India
		★ 2000 km High Speed Railways Network (HSR) by 2020
	National Highway Development Project	★ To meet the need for the provision and maintenance of National Highways network to global standards
		★ Improving more than 49,260 km of arterial routes of NH Network promote economic wellbeing and quality of life of the people
Efficiency	Fuel Economy Standards for cars	★ Binding fuel economy standards starting 2017
		★ Fuel Efficiency improvement in cars by 10 per cent in 2017
		★ 20 per cent in 2022 relative to 2010 levels
	Auto Fuel Policy	★ Phased implementation of Vehicle and Fuel Quality norms in the country

Sector	Policy/Plan	Highlights
Freight	Dedicated freight corridors	★ Double employment potential in five years (14.8 per cent CAGR)
		★ Triple industrial output in five years (24.57 per cent CAGR)
		★ Quadruple exports from the region in five years (31.95 per cent CAGR)

Source: Adapted from Shukla and Pathak (2016)

National Urban Transport Policies of India

One of the major initiatives that triggered increased attention to sustainable transport in cities was the formulation and adoption, in April 2006 (NUTP, 2006), of a National Urban Transport Policy. Under severe pressure for "doing something" about the deteriorating transport situation in the largest cities of the country, the Ministry of Urban Development decided to formulate a National Urban Transport Policy and, in April, 2003, set up a committee to prepare a draft (Agarwal, O.P. *et al.*, 2006). The objective of this policy is to plan for the people rather than vehicles by providing sustainable mobility and accessibility to all citizens to jobs, education, social services and recreation at affordable cost and within reasonable time (NUTP, 2006). The major elements of this Policy are the following (MoUD, 2014):

★ Incorporating urban transportation as an important parameter at the urban planning stage rather than being a consequential requirement.

★ Bringing about a more equitable allocation of road space with people, rather than vehicles, as its main focus

★ Public Transport should be citywide, safe, seamless, user friendly, reliable and should provide good ambience with well-behaved drivers and conductors.

★ Walk and cycle should become safe modes of Urban Transport.

★ Introducing Intelligent Transport Systems for traffic management

★ Addressing concerns of road safety and trauma response

★ Raising finances, through innovative mechanisms

★ Establishing institutional mechanisms for enhanced coordination in the planning and management of transport systems.

★ Building capacity (institutional and manpower) to plan for sustainable urban transport and establishing knowledge management system that would serve the needs of all urban transport professionals, such as planners, researchers, teachers, students, *etc.*

★ Support the principle that the Government provides the capital infrastructure, but the direct and indirect beneficiaries pay for the operating costs.

★ Innovative financing mechanisms, with greater involvement of private sector

* Encourage incentives that will facilitate the use of cleaner fuel and vehicle technologies, so that the pollution caused by motor vehicles gets reduced.

Approach

The estimate of the working group on Urban Transport for the 'National Transport Development Policy Committee' appointed by the Government of India to determine the role of Urban Transport in meeting transport requirements of the economy over the next two decades shows that the investment can be minimized by approximately 30 per cent by pro-actively encouraging sustainable practices. (MoUD, 2014)

Thus, a paradigm shift is needed in approach to Urban Transport with three key strategies, namely, 'Avoid, Shift and Improve' in transport planning as advocated by the Asian Development Bank in its draft 'Action Plan to Make Transport in Developing Countries more Climate Friendly' and reiterated by the Bellagio Declaration 8 in May 2009 (MoUD, 2014). This means, 'avoid' increase in demand for travel both by reducing the number and length of trips, promote a 'shift' from personal vehicles to other MRT and Non-Motorized Transport (NMT) modes to reduce energy demand and hence pollution in cities and 'improve' strategy on use of clean fuels and clean vehicle technology (MoUD, 2014).

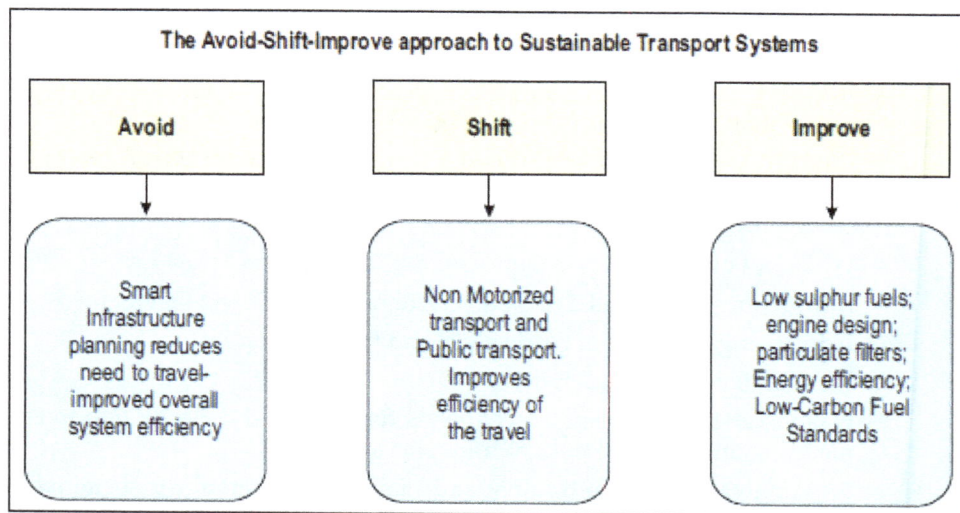

Figure 8.2: Avoid, Shift and Improve (ASI) System (Ramanathan V. *et al.*, 2014).

According to Ministry of Urban Development, Urban Transport Planning including Integrated Land Use and Transport Planning, Comprehensive Mobility Planning (CMP), Modal Mix and Priorities, Transit Oriented Development, Transportation Demand Management (TDM), Controlling the use of personal vehicles, Planning for Freight Traffic, Service Level Benchmarks (SLB) and Public participation increases the likelihood that actions taken or services provided by public agencies reflect the needs of people and are accepted/adopted by people easily.

National Urban Renewal Mission

Concurrently the National Urban Transport Policy was being discussed, a realisation was also emerging that the hitherto emphasis on rural development had led to a neglect of urban sector, which suffered from poor infrastructure. The occurrence of severe rains in Mumbai, in 2004, which virtually brought the commercial capital of India to a stand-still, helped highlight the neglect faced by urban areas (Agarwal, O.P. *et al.*, 2006). Also, realising that though less than 30 per cent of the population resided in urban areas, more than 60 per cent of the GDP came from such areas, the government decided that the time had come to correct the past neglect. It launched the "Jawaharlal Nehru National Urban Renewal Mission (JnNURM)" (JnNURM, 2005) in December 2005 committing substantial funds for investments in urban infrastructure. Speaking at the formal launch event of this Mission, the Minister for Urban Development said that this manifested recognition of the fact that urbanisation was irreversible and here to stay. This mission had the following features (Agarwal, O.P. *et al.*, 2006):

* The requirement that a city first prepares, after extensive public consultation, an overall City Development Plan (CDP), presenting a strategic vision of what the city wanted to be;

* A priority list of investments was required to be incorporated in the CDP so that investments are made with a clear priority in mind and not in an ad-hoc manner;

* Central Government's support by way of financial grants for investments in urban infrastructure ranging from 35 per cent of the project cost for the large cities which have the ability to raise resources on their own, to 50 per cent for medium sized cities and going up to 90 per cent for smaller, disadvantaged cities;

* Substantial funds earmarked for this purpose - the National Government committed an amount of Rs. 50,000 Crore (approximately $10 billion) to be spent over a period of 7 years, with the expectation that this would leverage at least a similar amount from State Governments, the private sector and financial institutions;

* Reforms in urban governance to enable financial sustainability of the physical assets created and their operation – essentially to ensure that the assets created would not need financial support on a continuous basis and would be able to recover their operating costs;

* Establishment of a Central Sanctioning and Monitoring Committee (CSMC) that would first scrutinize the CDPs and only thereafter approve individual project proposals for the Central Government's financial support.

The National Urban Renewal Mission, incorporate with the NUTP, provided the thrust needed for the changes that took place in seeking solutions to the problems of urban gridlock and poor air quality. JnNURM provided substantial funds and, armed with generous support from the National budget, cities were able to come up with ambitious plans for infrastructure development. Incorporated with this, the NUTP provided a framework and a direction for possible interventions towards sustainable transport.

India has launched two schemes to drive urban transformation and economic growth – The "Smart Cities Mission" and the "Atal Mission for Rejuvenation and Urban Transformation (AMRUT)".

The "Smart Cities Mission" will cover 100 cities with the objective of promoting area-based urban development initiatives that are replicable. The key focus of the Smart Cities Mission is to provide a decent quality of life to citizens through "Smart" solutions, enabled by technology applications, while maintaining a clean and sustainable environment.

The "Atal Mission for Rejuvenation and Urban Transformation (AMRUT)" will cover 500 cities. AMRUT seeks to provide basic services (*e.g.* water supply, sewerage, urban transport) to improve the quality life for all, especially the poor and the disadvantaged.

Both these missions have been designed based on learning's emerging from the JnNURM, the previous urban development scheme administered during 2007-2014.

National Mission on Sustainable Habitat

The National Mission on Sustainable Habitat (NMSH) predicts a framework to build urban resilience to climate change, by amalgamating adaptation and mitigation aspects into the urban planning process. The NMSH describes strategies to implement measures in the various sectors listed below (Venkataramani V. and Shivaranjani V., 2015):

* ★ Energy Efficiency
* ★ Urban Transport
* ★ Water Supply and Sewerage
* ★ Municipal Solid Waste Management
* ★ Urban Storm Water Management
* ★ Urban Planning

The major elements of the National Mission on Sustainable Habitat are the following:

* ★ Extending the existing Energy Conservation Building Code
* ★ A greater emphasis on urban waste management and recycling, including power production from waste
* ★ Strengthening the enforcement of automotive fuel economy standards and using pricing measures to encourage the purchase of efficient vehicles; and
* ★ Incentives for the use of public transportation.

National Policy on Biofuels

The National Policy on Biofuels focuses at mainstreaming of biofuels. The Policy will bring about accelerated development and promotion of the cultivation, production and use of biofuels to increasingly substitute petrol and diesel for transport and be used in stationary and other applications, while contributing to energy security, climate change mitigation, apart from creating new employment opportunities and leading to environmentally sustainable development (Government of India. 2008). Therefore, envisions a central role for it in the energy and transportation sectors of the country in coming decades. The Goal of the Policy is to ensure that a minimum level of biofuels become readily available in the market to meet the demand at any given time.

Salient Features (Basavaraj G., *et al.*, 2012)

★ An indicative target of 20 per cent blending of biofuels both for biodiesel and bioethanol by 2017

★ Biodiesel production from non-edible oilseeds on waste, degraded and marginal lands to be encouraged

★ A Minimum Support Price to be announced for farmers producing non-edible oilseeds used to produce biodiesel

★ Financial incentives for new and second generation biofuels, including a National Biofuel Fund

★ Biodiesel and bioethanol are likely to be brought under the ambit of "declared goods" by the Government to ensure the unrestricted movement of biofuels within and outside the states

★ Setting up a National Biofuel Coordination Committee under the Prime Minister for a broader policy perspective

★ Setting up a Biofuel Steering Committee under the Cabinet Secretary to oversee policy implementation.

National Electric Mobility Mission Plan

National Electric Mobility Mission Plan (NEMMP) 2020 was launched by the government of India in 2013. This plan is one of the most important and ambitious initiatives of the government to bring about transformational change in the automotive and transport landscape in the country.

Its objective is that 6-7 million of different types of electric and hybrid vehicles should be sold by 2020, which will result in 2.2-2.5 million tonnes of liquid fuel savings and a decrease of 1.3 – 1.5 per cent in carbon dioxide emissions. It has envisaged a comprehensive scheme, covering all aspects of electric mobility:

★ Incentive to facilitate acquisition of hybrid or electric vehicles;

★ Promoting R&D in technology including battery technology, battery management system, testing infrastructure, power electronics, motors and ensuring industry participation in the same

★ Power and charging infrastructure;

★ Supply-side management;

★ Encouragement of retrofitting of old vehicles with hybrid kits

Auto Fuel Policy

The National Auto Fuel Policy announced by the Petroleum Minister, Mr. Ram Naik on October 6, 2003 envisaged a phased programme for introducing vehicular emission norms in the country by 2010. The policy sought to improve the fuel quality and vehicular engine specifications. It had proposed that liquid fuels remain the main auto fuel throughout the country and suggested the use of CNG and LPG in cities affected by higher pollution levels to enable vehicle owners have the choice of the fuel and technology combination.

Issues and Challenges

★ **India's roads are congested and of poor quality.** Lane capacity is low - most national highways have two lanes or less. A quarter of all India's highways are congested. Many roads are of poor quality and road maintenance remains under-funded - only around one-third of maintenance needs are met. This leads to the deterioration of roads and high transport costs for users. (MoRT and H – WTTC)

★ Governance of urban transport is still highly fragmented and urban transport actions are managed by a multitude of agencies in any city. Often these agencies report to different levels of government. Such fragmentation leads to uncoordinated planning with sub-optimal outcomes (Agarwal, O.P. *et al.*, 2006).

★ In metropolitans there is a lack of fast and adequate public transport system. This inadequacy leads to escalation of personal transport (own vehicles) which puts extra pressure on roads and cause jams and accidents. Further, hilly and remote areas lack all-weather transport facility. (Economics Discussion)

★ Integrated land use and transport planning has not been institutionalised. Although cities have been required to prepare Comprehensive Mobility Plans (CMPs), these have, at best, tried to take into account multiple transport sub-systems, without really integrating them as part of a larger city transport system network. In any case, land use and transport plans have not been integrated (Agarwal, O.P. *et al.*, 2006).

★ National Urban Transport Policy is in place, but it lacks legal backing due to which non-compliance with the policy cannot be penalised. However, the agencies are directly responsible for urban transport sector standard, but those less directly responsible for it do not give it the seriousness it deserves.

★ Policies for more and more road construction have clearly failed to cope with ever increasing demand from rapid motorization, resulting in a vicious circle as depicted in Figure 3 (Banister, D. 2008). This cycle shows

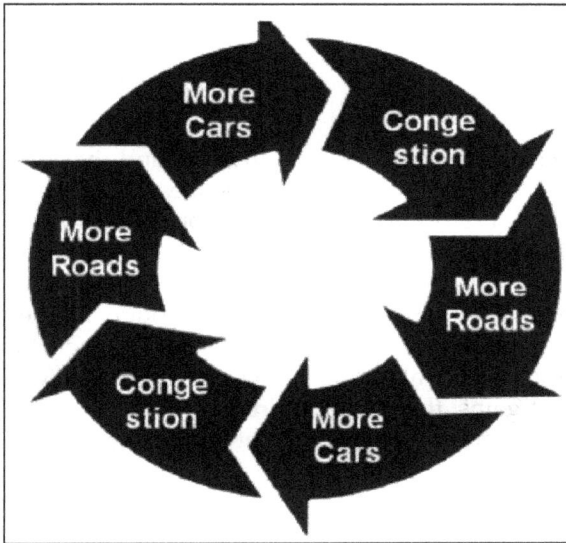

Figure 8.3: Vicious Circle of Car-Oriented Transport Development [Banister, D. (2008)].

how the increase of infrastructure to alleviate travel demand will have apparently positive consequences in the short term, but some months later there will be a much greater congestion than before, thus increasing the problem rather than solving it. (Banister, D. 2008)

★ Though the JnNURM sought to offer funds for initiatives that complied with the National Urban Transport Policy, a substantial portion was allocated to road widening projects –those that were out of line with the NUTP. Almost 60 per cent of the total funds allocated to the Urban Transport Sector were towards projects for building roads, flyovers, and rail over bridges. (Agarwal, O.P. *et al.*, 2006)

★ There have been some efforts towards promoting alternative fuels, but these have been few and will not be able to go a long way. On a direction from the Supreme Court of India, Delhi was compelled to convert all its buses and para-transit to CNG. Yet other cities have not replicated this. A few small areas do have electric vehicles and an Electric Vehicle Task Force has also been set up under the Ministry of Heavy Industry. However, these have not led to much visible progress towards alternative fuels, especially when the consumption of petroleum fuels and the impact on the nation's foreign exchange is so large. Thus an urgent focus on investment for electrification of the public bus system is vital. Although initial investments may be large, plans of technology transfer and indigenisation of the technology would make this effort sustainable (Agarwal, O.P. *et al.*, 2006).

★ All the country's high-density rail corridors face severe capacity constraints. Also, freight transportation costs by rail are much higher than in most countries as freight tariffs in India have been kept high to subsidize passenger traffic.

Conclusions

It has become clear that massive urbanisation and the transport problems it presents will be one of the most important challenges facing India in the future. The linkage between transportation and GHG emissions makes it imperative to devise strategies that limit the emissions and simultaneously make the urban areas resilient to climate change. The review of the literature observed that the Government of India has taken important steps to meet the challenge through a variety of mechanisms primarily through the adoption of a National Urban Transport Policy, the launch of the National Urban Renewal Mission, smart cites and AMRUT. These have led to several important gains, the most important of which has been the realisation that public transport improvement and not road capacity enhancement is the way forward. However, several challenges still remain. The way forward needs to emphasise a comprehensive and coordinated approach rather than a fancy for high cost facilities. It needs to emphasise governance structures that enable comprehensive planning and coordinated implementation. It needs to work towards innovative financing and alternative fuels. It needs to emphasise that good urban transport planning has to be "people" focused rather than "Engineering" focused.

References

1. IPCC (2007). Climate Change 2007: Impacts, Adaptation and Vulnerability. Contribution of Working Group II to the Fourth Assessment Report of the Intergovernmental Panel on Climate Change, M.L. Parry, O.F. Canziani, J.P. Palutikof, P.J. van der Linden and C.E. Hanson, Eds., Cambridge University Press, Cambridge, UK, 976 p.

2. IPCC (2001). Climate Change 2001: The Scientific Basis. Contribution of Working Group I to the Third Assessment Report of the Intergovernmental Panel on Climate Change. Houghton, J.T., Y. Ding, D.J. Griggs, M. Noguer, P.J. van der Linden, X. Dai, K. Maskell, and C.A. Johnson (eds.). Cambridge University Press, Cambridge, United Kingdom and New York, NY, USA. pp. 881.

3. IEA (2015). CO_2 Emissions from Fuel Combustion Highlights,10 p.

4. Mittra S. (2016). Deconstructing the Climate-Conflict Nexus in Urbanising India. ORF Issue Brief, Issue No. 148. pp. 1-4.

5. Seto, Karen C., Burak Güneralp, and Lucy R. Hutyra (2012). Global forecasts of urban expansion to 2030 and direct impacts on biodiversity and carbon pools. *Proceedings of the National Academy of Sciences* 109, No. 40. pp. 16083-16088.

6. NAPCC, 2008. http://pmindia.nic.in/Pg01-52.pdf (assessed on 11 Oct 2008).

7. http://www.economist.com/news/specialreport/21651324-indias-future-urban-let-there-be-concrete (23rd May 2015)

8. Estimates from the CASI report: https://casi.sas.upenn.edu/india per cent E2 per cent 80 per cent 99s-urban-future/india per cent E2 per cent 80 per cent 99s-urban-future

9. Trivedi, P. (2015). "Modi at COP 21: India will negotiate hard to keep its carbon space". Indian Express.

10. McGranahan, Gordon, M., Balk, D., and Anderson, B., 2007. The rising tide: assessing the risks of climate change and human settlements in low elevation coastal zones, Environment and Urbanization, International Institute for Environment and Development (IIED). Vol 19, No 1, April, pp. 17–37.

11. Census of India (2011). Provisional Population Total, Government of India Press, New Delhi.

12. Singh, K.N. (1 January 1978). Urban Development In India. Abhinav Publications. ISBN 978-81-7017-080-8. Retrieved 13 June 2012.

13. Ministry of Road Transport and Highways (MoRT and H) (2012). Road Transport Yearbook (2009-10 and 2010-11), Transport Research Wing, Ministry of Road Transport and Highways, Government of India, New Delhi.

14. NTDPC (2013). Report of the Working Group on Urban Transport, Ministry of Urban Development.

15. Singh, 2006. Future mobility in India: Implications for energy demand and CO_2 emission, Transport Policy pp. 13: 398–412.

16. Ramanathan V. *et al.*, 2014: India California Air Pollution Mitigation Program: Options to reduce road transport pollution in India. The Energy and Resources Institute in collaboration with the University of California at San Diego and the California Air Resources Board.

17. Ministry of Petroleum and Natural Gas (MoPNG) (2013). Petroleum and Natural Gas Statistics 2012-2013. New Delhi: MoPNG, Government of India.

18. Ministry of Environment and Forests (MoEF) (2010). India: Greenhouse gas emissions 2007. New Delhi: Indian Network for Climate Change Assessment (INCCA), MoEF. Government of India.

19. Shukla, P. R., and Pathak, M. (2016). forthcoming. Low-carbon transport in India: Assessment of best practice case studies, Chapter 8. In: Enabling Asia to Stabilize Climate. Springer.

20. IEA (International Energy Agency) (2012). Understanding Energy Challenges in India: Policies, Players and Issues, OECD/IEA, Paris.

21. National Urban Transport Policy (2006). Retrieved June 20, 2014, from http://urbanindia.nic.in/policies/TransportPolicy.pdf

22. Ministry of Urban Development (2011). Retrieved June 24, 2014, from http://moud.gov.in/

23. Agarwal, O.P. *et al.* (2006). Review of urban transport in India. Institute of Urban Transport(India).

24. Ministry of Urban Development (MoUD) (2014). National Urban Transport Policy, 2014. New Delhi: MoUD, Government of India. Details available at: http://itdp.in/wp-content/uploads/2014/11/NUTP-2014.pdf

25. Venkataramani V. and Shivaranjani V. (2015). Mission Brief prepared as part of the Study: Implementation of the National Action Plan on Climate Change (NAPCC) - Progress and Evaluation. Centre for Development Finance (CDF), IFMR LEAD.

26. Government of India (2008). National Policy on Biofuels, Ministry of New and Renewable Energy. Government of India.

27. Banister, D. (2008). "The Sustainable Mobility Paradigm," Transport policy, 15(2), 73-80.

28. Basavaraj G., *et al.* (2012). A Review of the National Biofuel Policy in India: A critique of the Need to Promote Alternative Feedstocks, Working Paper Series no. 34, RP – Markets, Institutions and Policies 2012.

29. Divya Sharma and Sanjay Tomar (2010). Mainstreaming climate change adaptation in Indian cities. International Institute for Environment and Development (IIED). 451 Vol 22(2): 451-465. DOI:10.1177/0956247810377390.

30. Eleventh (Planning Commission). http://planningcommission.nic.in/plans/planrel/fiveyr/11th/11_v3/11v3_ch9.pdf

31. Dhar, S., Pathak, M., and Shukla, P. (2015). Transport Scenarios for India: Harmonising Development and Climate Benefits. UNEP DTU Partnership.

32. Ministry of Road Transport and Highways (MoRT and H) – WTTC India, http://www.wttcii.org/pdf/File per cent 206.pdf

33. http://www.economicsdiscussion.net/articles/main-problems-of-transport-development-in-india/2183

34. Puchera, J., *et al.* (2005). Urban transport crisis in India. Transport Policy 12 (2005) 185–198

35. Verma, A., *et al.* (2015). Urban Transport Policies in India in context to Climate Change: An International Perspective. 10th Annual Conference of Knowledge Forum

Chapter 9

Green Rating for Integrated Habitat Assessment (GRIHA): An Indian Model for Building Sustainability

Rashmi Srivastava

Research Associate,
Centre for Science and Technology of the Non Aligned and
Other Developing Countries (NAM S&T Centre), New Delhi, India
E-mail: rashmisri25@gmail.com

Abstract

Buildings are one of the significant consumers of energy. Approximately, 30 – 40 per cent of the global Greenhouse Gases (GHGs) emissions result from activities, which are directly or indirectly, engaged in buildings and construction. With increase in urbanization, demand for energy in the building sector has forced researchers to delve into newer ideas for energy conservation. In this paper, an effort has been made to deliberate on Green Rating for Integrated Habitat Assessment (GRIHA) to assess energy efficient designs for multi – dimensional building spaces. GRIHA is one such instrument, conceived for establishing and accomplishing sustainable habitats for every being, across the country.

Keywords: Green Rating for Integrated Habitat Assessment (GRIHA), Sustainable habitat, Greenhouse Gases (GHGs), Urbanization, Multi-dimensional building spaces.

Introduction

> *We shape our buildings; thereafter they shape us.*
>
> *~Winston Churchill*

The ancient Indic heritage is widely known for its spacious divine home for the Environmental ethos. Its various infrastructures such as forts, palaces, tombs, reservoirs, roadways, canals *etc.* were in absolute unison with the nature and the occupants. They were more often concerned in less wastage of the resources and health of the dwellers. However, over the last decade, the buildings have become the icon of modernity and advancement. Their un-planned construction across the globe, have been increased in manifold sectors *viz.*, roads, railways, households, schools, institutions, hospitals *etc.* India is planning to invest about US $ 1 trillion in infrastructure development, over next five years (World Bank Report, 2015).

Future infrastructure alone, mostly in the urban global South, would require an estimated third of the remaining carbon budget (Muller, *et al.,* 2014). This increase in demand of energy is borne out of the country's efforts to industralise, urbanise, remove poverty and provide higher standard of living to its population. This industralisation will require strengthening of manufacture sector, which will ensure in constructing reliable infrastructures for its citizens. In the light of what was achieved on 12th December, 2015 at COP 21 in Paris, it is evident that the agreement allows India enough room to plan and undertake an industrialization process (which is largely predicted) to proliferate on the basis of two sources of energy, namely, coal and renewable energy sector. However, there are several trends *viz.*, growth, urbanization, technology, globalisation *etc.* (illustrated in Figure 9.1) which will continue to shape the infrastructure agenda of the future.

Globally, it has been found that more people dwell in urban areas than in rural areas, with 54 per cent of the world's population residing in urban centres (UN, 2014). The proportion of rural and urban population, in a rapidly urbanizing country like India, Nigeria and China, has always been in a debate across the globe.

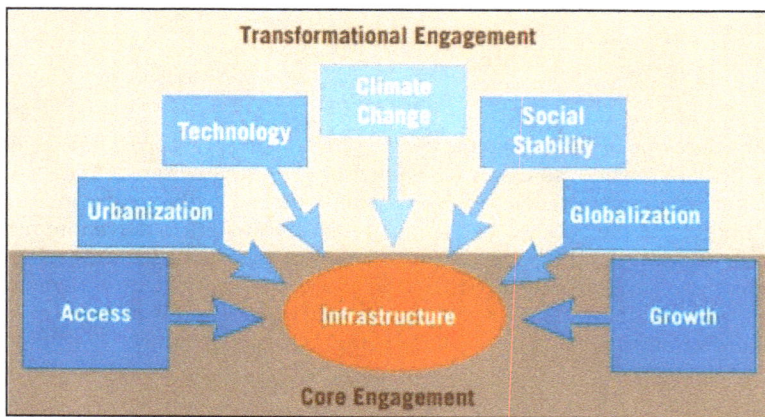

Figure 9.1: Flowchart Showing different Trends that Shape the Infrastructure Growth (*Source*: www.grihaindia.org).

According to UN population projections, of all the countries in the world, India is expected to experience the largest increase in urban population with 497 million persons being added to the existing urban population by 2050. But, quite distinctively, this transition is not only due to natural increase or rural to urban migration but it is also due to 'in-situ urbanisation' taking place in two ways:

reclassification of villages into towns with increasing population and engulfing sub-urban villages by the expanding desire or needs of towns and cities (Kornegay *et al.*, 2013; Pradhan, 2012). In this process, economic and demographic changes are being evident with people shifting from primary to secondary and tertiary sectors (Lahoti and Swaminathan, 2013). This urban transition, anticipated to usher India into a new growth trajectory, can pose threats to economic, social, environmental sectors, if not managed well (Aijaz, 2015). The population immigration from rural to urban areas to avail air, water, food, land *etc.* successively increases the construction of number of infrastructures for accommodation, schools, offices, hospitals *etc.*, which will in turn, drastically increase the requirement of energy and resources.

From the above discussion, therefore, it can be deduced that the unprecedented rate of urbanisation has increased environmental vulnerabilities to climate change in 21st century. First, since Fourth Assessment Report (AR4) of the IPCC, it was ascertained that GHG emissions have continued to grow globally and have reached 49.5 billion tonnes (giga tonnes or Gt) of CO_2 eq in the year 2010, higher than any level prior to that date, with an uncertainty estimate at ±10 per cent for the 90 per cent confidence interval. Further, AR5 of IPCC mentioned that the current trajectory of global annual and cumulative emissions of GHGs is inconsistent with widely discussed goals of limiting global warming at 1.5 to 2.0 degree Celsius above the pre-industrial level. In this regard, the meeting of world leaders at COP 21in 2015 was being seen by the different international community as a landmark opportunity to seize the momentum for influential climate action. Implementation of the Paris Agreement has raised new concerns and challenges regarding climate equity, technology, finance and mitigation. This paper, however, deconstruct on how building sector can realise and combat climatic aberrations with much seriousness and vigour that will help in mitigating GHGs emissions sustainably.

Buildings are the most apprehended and evident sectors of unprecedented urbanisation and climatic changes. It has been estimated that construction of buildings emits 30-40 per cent of GHGs and generates 60 per cent of waste by its related activities. Given the massive growth in new construction, it has also been observed that infrastructure consumes up to 40 per cent of all energy during its construction, operation and maintenance. Thereof, creation of green buildings globally plays a crucial role, not only in mitigating climatic changes but also in achieving 17 Sustainable Development Goals (SDGs) with 169 targets; given prime importance to poverty eradication, and endorse within them infrastructure development and manufacturing-led industrialisation.

With India, turning out to be the *third* world's largest construction market in the world by 2025, adding 11.5 million homes a year to become a $1 trillion a year market (Sen, 2013) it is imperative to innovate, assimilate and diffuse strategies or methods which amplify footprints of the Green Buildings across the Nation. Green Rating for Integrated Habitat Assessment (GRIHA) is one such indigenous green building rating system developed for the Indian construction scenario. It has been conceived by The Energy and Resources Institute (TERI) and jointly developed by Ministry of New and Renewable Energy (MNRE) for deconstructing and assessing buildings in India.

GRIHA is adopted as the National Rating System (NRS) under the MNRE on 1 November, 2007. It takes into account the provisions of the National Building Code 2005, The Energy Conservation Building Code 2007 by Bureau of Energy Efficiency (BEE), IS codes, local bye-laws and other local standards laws; for developing a sustainable approach to the built environment. It brings together Indian's traditional architecture and modern technology to craft Green buildings.

Table 9.1 entails success of GRIHA, since 2007-2015.

Table 9.1: Achievements of GRIHA, from 2007-2015 [7]

Sl.No.	Heads	GRIHA
1.	Inception Year	2007
2.	Number of Projects Registered totaling to over 230 million square feet	650
3.	Total Buildings Rated	30
4.	Cumulative Annual Energy Consumption Reduction	74, 000MWh
5.	Installation of Renewable Energy	14.5MWp
6.	Professionals Trained	10,000+
7.	Accredited Professionals	466+

By its acronym, GRIHA - a Sanskrit word - means *'Abode'*, is a tool for evaluating the environmental performance of a building holistically over its entire life cycle, thereby enabling in deriving a definitive standard for what constitutes a 'green building'. By its already set strategies, GRIHA ensures that building must use Energy and Water optimally and generates minimum waste in its period of construction. Above all, it certifies the well-being of each individual *i.e.* from labourers to the dwellers in the buildings. The system, by its qualitative and quantitative assessment criteria, is able to 'rate' a building on the degree of its 'greenness'. The rating is applied to new and existing building stock of varied functions—commercial, institutional, and residential sectors *i.e.* it is suitable for different climatic zones of the country.

As a matter of fact, buildings play significant role in our daily lives. Be it from accommodation to offices or schools/restaurants or even its construction to maintenance *etc*. All our activities directly or indirectly revolve around buildings. Apart from the basic functions of buildings, as discussed above, GRIHA broadly defines the two fundamental physiological 'comforts' to its occupants, namely:

★ Visual comfort – the ability of the inhabitants to see clearly for carrying out their daily domestic or official tasks.

★ Thermal comfort – the ability of the building to regulate temperature for its occupants during winter and summer seasons, thereby enabling the state of homeostasis.

The above categorised comforts, however, can be acquired by natural and artificial means *viz.*, by using sun-light, natural winds, evaporation, trees, electric lighting, ACs *etc.* By examining the building closely, GRIHA ensures that its occupants must attain these 'comforts', simultaneously looking forward to accomplish environmental equity and sustainability.

Supplement to this, there are 3 forms of certification provided by GRIHA, namely: [7]

1. SVA GRIHA Certification for 100 square meters to 2,499 square meter of built-up area.

2. GRIHA for new construction Certification for 2,500 square meters to 1, 50,000 square meter.

3. GRIHA LD is for the site areas exceeding to 50 hectares or built-up area more than 1, 50,000 square meters.

Hitherto, it has been found that GRIHA has accorded 900 projects with 36 million sq. meters (approximately) [8].

Benefits of Griha

As a rating tool, GRIHA, aids people to evaluate their building performances against certain nationally acceptable environmental benchmarks. On a larger scale, it benefits the community with the improvement in the surroundings by reducing GHGs emission, energy consumption and stress on natural resources. The outcomes of different research have showed that most of the buildings projected to be standing in 2030 in India are yet to be constructed. Therefore, to promote sustainability by assessing different Green buildings, GRIHA came into practice and has the advantages given on next page [7].

The Mechanism Involved

As discussed above, GRIHA attempts to quantify different aspects of the Environment, such as energy consumption, waste generation, renewable energy adoption *etc.* The rating process was initially comprised of 34 criteria which are now revised to 31 in 2015. This amendment was imperative so as to bring about the *better understanding and implementation of current market scenario and formulate new benchmarks to induce the market further towards sustainability.* This paper, briefly illustrates the revised criteria formulated and implemented by GRIHA and MNRE. These criteria assist in scaling buildings on a scale of 0-100 (+4 for innovations) points with a minimum of 50 points required for a building to be rated under GRIHA. Rating is denoted in terms giving stars from *one* to *five* depending upon the total score gained by the buildings, wherein *five* denotes the infrastructure which is most sustainable into the environment and *one* depicts its least sustainability. It should also be noted that in *GRIHA Energy Performance Index* (EPI - total energy consumed in a building for a unit built-up area for one year) *is evaluated in kWh/sq.m./annum.*

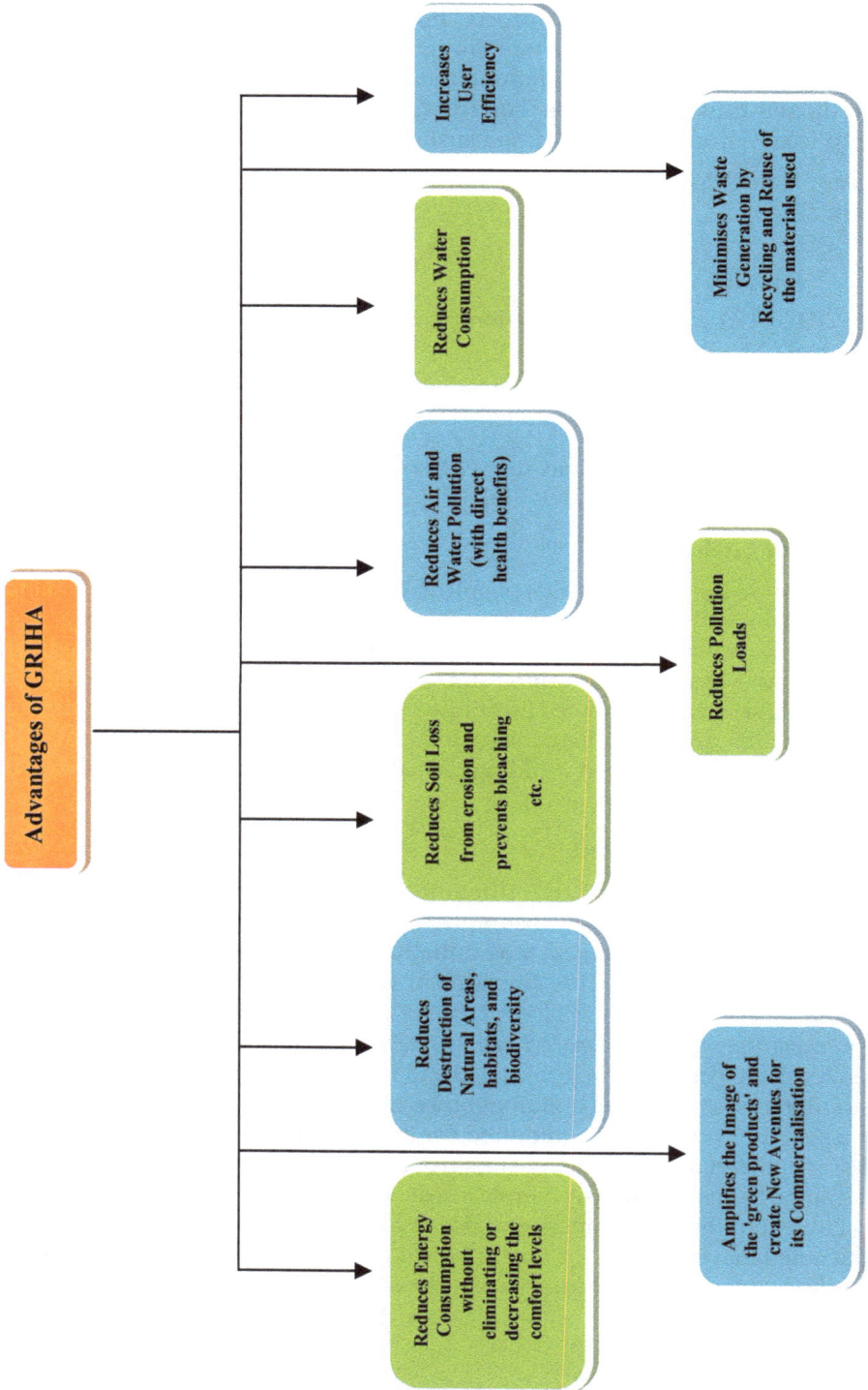

Advantages of GRIHA

- Increases User Efficiency
- Reduces Water Consumption
- Reduces Air and Water Pollution (with direct health benefits)
- Reduces Soil Loss from erosion and prevents bleaching etc.
- Reduces Destruction of Natural Areas, habitats, and biodiversity
- Reduces Energy Consumption without eliminating or decreasing the comfort levels

- Minimises Waste Generation by Recycling and Reuse of the materials used
- Reduces Pollution Loads
- Amplifies the Image of the 'green products' and create New Avenues for its Commercialisation

Table 9.2: Revised Criteria Adopted by GRIHA for Assessing Green Buildings [7]

Sections	Criteria Number	Criteria Names	Points Allocated
Site Planning	1	Site Selection	1 (Partly Mandatory)
	2	Low Impact Design	4 (Partly Mandatory)
	3	Design to mitigate Urban Heat Island Effect (UHIE)	2
	4	Site Imperviousness Factor	1 (Mandatory)
Construction Management	5	Air and Water Pollution control	1
	6	Preserve and protect landscape during construction	4
	7	Construction Management Practices	4
Energy	8	Energy Efficiency	13
	9	Renewable energy utilization	07
	10	Zero ODP materials	0
Occupant Comfort and Well Being	11	Achieving indoor comfort requirements (visual/thermal/acoustic)	6
	12	Maintaining good Indoor Air Quality (IAQ)	4
	13	Use of low VOC paints/adhesives/sealants in building interiors	2
Water	14	Use of low-flow fixtures and system	4 (Mandatory)
	15	Reducing landscape water demand	4
	16	Water Quality	2
	17	On-site Water reuse	5
	18	Rain Water recharge	2
Sustainable Building Materials	19	Utilization of BIS recommended waste materials in building structures	6
	20	Reduction in embodied energy of building structure	4
	21	Use of low-environmental impact materials in building interiors	4
Solid Waste Management	22	Avoid post-construction landfill	4
	23	Treat organic waste on	2
Socio-Economic Strategies	24	Labour safety and sanitation	1
	25	Design for Universal Accessibility	2
	26	Dedicated facilities for service staff	2
	27	Increase in environmental awareness	1
	28	Smart metering and monitoring	8
Performance Monitoring and Validation	29	Operation, Maintenance Protocols	0
	30	Performance Assessment for Final Rating	0
	31	Innovation	4
		Total 100	

Further, on the basis of the above criteria, the following scale of rating system for green infrastructures was speculated.

Table 9.3: Modified Star Rating Strategy Adopted by GRIHA in 2015 [7]

Points Scored	Rating
25-40	*
41–55	**
56-70	***
71-85	****
86 or more	*****

On a bigger picture, GRIHA enables assessing and thereof reducing GHGs emissions into the environment. By scaling various buildings on the grounds of necessary environmental benchmarks, GRIHA is facilitating carbon footprint reduction to foster sustainability by eliminating the staggering demand of energy, water and other inputs required in building infrastructures.

Case Study

The following *two* case studies briefly outlay the implementation and benefits of GRIHA at different locations of India [10].

Table 9.4: Two Case Studies

Heads	Case Study I	Case Study II
Name	COAL INDIA LIMITED OFFICE BUILDING	INDIAN INSTITUTE OF TECHNOLOGY
Location	Kolkata, West Bengal	Gandhi-Nagar, near Village Palaj, Gujarat
Site Area	20,235 m²	162.7 hectares/16, 26,838 m²
Built up Area	25,519.25 m²	5, 51,965 m²
Air-conditioned Area	13,892 m²	Not mentioned
Non Air-conditioned Area	11,627.25 m²	Not mentioned
Energy Consumption Reduction	37.9 per cent reductions in energy consumption compared to GRIHA benchmark	37 per cent below GRIHA LD base case
Energy Performance Index	86.88 kWh/m²/year	Not mentioned
Renewable Energy	Rated capacity of solar PV installed on site is 140 kWp	500 kWp (Installation proposed)
GRIHA provisional rating	4 stars	5 stars
Year of completion	2014	Not mentioned
Water Reduction	Not mentioned	30 per cent below GRIHA LD base case
Organic Waste Reduction	100 per cent	100 per cent

Figure 9..3: Coal India Limited Office Building, Kolkata (Case Study I).

Figure 9.4: Indian Institute of Technology, Gandhinagar (Case Study II) [7].

The following different approaches were advocated by Integrated Design Team (IDT) to minimize environmental degradation at these two sites; and GRIHA rating was assessed [7]:

Table 9.5: Comparison of the 2 Cases

Environmental Parameters	Characteristics	
	Case Study I (Coal India Limited Office Building, Kolkata)	Case Study II (Indian Institute of Technology, Gujarat)
Sustainable Site Planning	1. Existing contours were referred for constructing this building. 2. Light wells and courtyards are constructed to capture and harvest daylight. 3. To minimise soil erosion, building is constructed in appropriate season.	1. Over 79 per cent of total area under existing tree clusters, lakes and ravines are preserved. 2. The East-West directions (buffer Spaces) are used to build services and toilets. 3. Storm Water Management is achieved by reducing peak run-off quantity.
Energy Management and Renewable Energy Induction	1. 82 per cent of the total living-area captures day light (by constructing Light Wells, Court Yards and increasing External Shading and Efficient glazing) as prescribed by National Building Code (NBC). 2. For thermal management, following steps are undertaken: »» To reduce cooling load in AC spaces, fly ash brick in envelopes are used. »» To dehumidify air water cooled chillers are operated (as recommended in Energy Conservation Building Code). 3. For meeting daily energy requirements 140kWp solar panel is used. 4. 85 per cent of the total interior flooring is constructed with low-energy materials *viz.* granite stone, sandstone, vitrified tiles, hardonite tiles and with recycled content).	1. Buildings are 36 per cent more energy efficient by day lightening and low energy cooling as per the NBC norms. 2. Details are Unavailable 3. A 200kWp Grid Interactive Rooftop Solar PV Project is installed to harness solar energy for practical applications. 4. Details are Unavailable
Water Management	1. By installing 'Low Flow Fixtures', the water consumption is reduced by 78 per cent. 2. Use of premixed plaster and gunny bags for curing columns helped in reduction of water consumption, close to 50.17 per cent, while constructing the building.	1. By using rainwater, treated waste water and low flow fixtures, they have planned to minimize their annual water demand by 30 per cent. 2. Waste Water is recycled and reused by incorporating Decentralised Waste Water Treatment Systems (DEWATS)

Add-On Features of Griha for *Sustainable Habitat*

In 2017 GRIHA has further added many feathers in its cap. On 9th February, 2017 GRIHA Council signed a Memorandum of Understanding (MoU) with IREO for implementation of the Green Building concepts and the GRIHA Rating variants for IREO projects. IREO has already registered approximately 8 million square feet building footprint with GRIHA Council and is now in the process of registering an additional footprint of 2 million square feet [7]

Under this collaboration, IREO is committed to obtain GRIHA rating for its significant upcoming and future projects. As part of the MoU, GRIHA Council will aid IREO in setting up a project implementation unit for executing GRIHA Certification. Besides this, it will also work with IREO to maximize awareness and outreach on green initiatives through publicity campaign, marketing and collaterals. The MoU also aimed in establishing a review committee to monitor and provide guidance for smooth implementation of the green building initiatives.

Further, with the theme of 'Transforming Habitats', GRIHA team on 27th February, 2017 in the India Habitat Centre, New Delhi launched life size model of a green building replica entitled, 'SPARSH – GRIHA4ALL'. It basically deliberates on the linkages between urban governance and novel innovation capacities to transform and shape the future of our sustainable cities. The area of Green building mock-up was 36.55 square metre.

Figure 9.5: Outside View of SPARSH GRIHA4ALL [11].

The main objective behind the creation of SPARSH was to craft new infrastructures by utilizing innovative and efficient resources along with the inimitable and futuristic work. According to the GRIHA Council, "SPARSH is an effort to bring an economically viable and structurally sound green building option

In *SPARSH* #GRIHA4All, all fenestrations are made of recycled glass and have a very high thermal resistance property

In *SPARSH* #GRIHA4All, skylights are integrated in the design to reduce artificial lighting

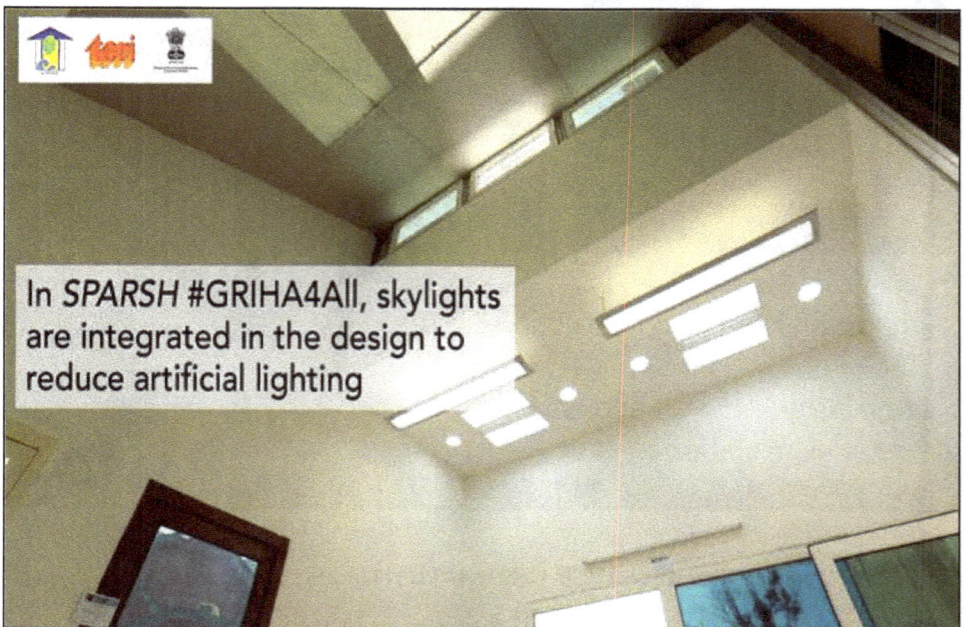

Figure 9.6: Inside view of SPARSH GRIHA [11].

to the masses. The cost of building a structure of any size on the SPARSH platform would be much less than the cost of traditional building of an equivalent size with similar amenities" [12].

The following discussion illustrates key features of the SPARSH Griha4All [13]:

* ★ Building envelope made on recyclable steel construction technique made by Galvalume alloy (Zi+Al+Si)
* ★ A Solar PV panel that supplies renewable energy to the structure
* ★ Building blocks that boast acoustic and thermal insulation
* ★ Fenestration made of recycled glass with a very high thermal resistance
* ★ State-of-the-art energy efficient lighting and appliances
* ★ Integrated operable sky-lighting to reduce artificial lighting
* ★ External walls made-up of cement fiber boards and gypsum boards using the dry wall technique
* ★ Internal and external flooring built out of construction and demolition waste
* ★ Post-industrial waste
* ★ Reducing the loads of landfill by 50 per cent
* ★ Water efficient bathroom fixtures reduce the water consumption of the structure by more than 50 per cent

Conclusions

Construction of Green Buildings is one such initiative adopted by India, to address and resolve urban-environmental challenges for constructing, operating and maintaining different infrastructures on an urban landscape. With India, turning out to be the *third* world's largest construction market in the world by 2025, it is imperative to innovate, assimilate and diffuse strategies or methods which amplify footprints of the Green Buildings across the Nation. However, the statistics reveals that the current green building infrastructure accounts for 3 per cent of 25 billion square ft., which is anticipated to reach 100 billion square ft. by 2030. Therefore, approaches outlined and enforced via GRIHA is quite noteworthy to aid the Nation in accomplishing sustainability beyond 2017 from the spectrum of Green Infrastructures.

References

1. Aijaz Rumi (2015): Global Policy, New Delhi: Observer Research Foundation Revi, *et al.,* Urban Areas. Climate Change 2014: Impacts, Adaptation, and Vulnerability. Part A: Global and Sectoral Aspects. Contribution of Working Group II to the Fifth Assessment Report of the Intergovernmental Panel on Climate Change. C.B. Field, V.R. Barros, D.J. Dokken, *et al.,* Cambridge, United Kingdom, and New York, USA, Cambridge University Press, 2014, pp. 535-612.

2. D.B. Muller, *et al.* (2013). Carbon Emissions of Infrastructure Development', Environmental Science and Technology 47(20): 11739-11746.

3. Kornegay Jr., Francis A. and Narnia Bohler-Muller (2013): Laying the BRICS of a New Global Order, Pretoria: Africa Institute of South Africa.

4. Lahoti, Rahul and Hema Swaminathan (2013): Economic Growth and Female Labour Force Participation in India, Working Paper No. 414, Bangalore: Indian Institute of Management. Available at: http://www.iimb.ernet.in/research/sites/default/files/WP per cent 20No. per cent 20414_0.pdf.[Accessed 01st February, 2017]

5. Pradhan, K.C. (2012): Unacknowledged Urbanisation: The New Census Towns of India', CPR Urban Working Paper 2, New Delhi: Centre for Policy Research. Available at: http://www.cprindia.org/sites/ default/files/ Unacknowledged per cent 20Urbanisation_CPR per cent 20Working per cent 20 Paper_KCPradhan_0.pdf.[Accessed 03rd February, 2017]

6. Websites www.grihaindia.org. [Accessed on 25th October, 2016]

7. http://www.grihaindia.org/?t=Green_Rating_for_Integrated_Habitat_Assessment# and library

8. http://www.grihaindia.org/files/GRIHA_V2015_May2016.pdf [Accessed on 02nd November, 2016]

9. http://www.grihaindia.org/images/casestudies/pdf/coal-India-Kolkata.pdf, http://www.grihaindia.org/images/casestudies/pdf/masterplan_casestudy_IIT_Gandhinagar.pdf [Accessed on 3rd November, 2016].

10. www.financialexpress.com [Accessed on 26th March, 2017]

11. http://www.business-standard.com/article/news-ians/green-building-s-life-size-replica-unveiled-117022800015_1.html [Accessed on 25th March, 2017]

12. http://www.teriin.org/files/GRIHA-SPARSH-press-release.pdf [Accessed on 26th March, 2017]

Part II

R&D in Energy

Chapter 10

Influence of Weather Sensitivity on Electricity Consumption in Cotonou, and Abidjan, Two Coastal Megacities in Western Africa

Kondi Akara Ghafi[1], Arona Diedhiou[2] and Benoit Hingray[3]

[1]*Excellence Center for Climate Change,*
Biodiversity and Sustainable Agriculture (CCBAD)
Université Félix Houphouët-Boigny (UFHB), Abidjan, Cote d'Ivoire
E-mail: vickykondi@gmail.com
[2]*Institut de Recherche pour le Développement (IRD), Grenoble, France*
E-mail: arona.diedhiou@ird.fr
Laboratoire de Physique de l'Atmosphère et Mécanique des fluides (LAPA-MF),
Abidjan, Cote d'Ivoire
E-mail: arona.diedhiou@gmail.com
[3]*Université Grenoble-Alpes, Laboratoire d'étude des Transferts*
en Hydrologie et Environnement (LTHE), Grenoble, France
E-mail: benoit.hingray@univ-grenoble-alpes.fr

Abstract

To provide insights on the possible climate change effects on the electricity demand in West African Cities, this study assesses the effects of weather variability on the electricity consumption. For this, the correlation between consumption and Cooling-Degree-Days (CDD) estimated from different temperature indices, including humidex and heat index was investigated. Monthly electricity consumption data from 1990 to 2015 for Abidjan (Côte d'Ivoire) and from 2000 to 2015 for Cotonou (Benin) were collected from national electricity services. The observation weather data were also collected from 1980 to 2014 from the meteorological synoptic stations for both cities. Statistical analysis was done by using the classical multiplicative decomposition of consumption data over sub-periods where the consumption was estimated to have a relatively homogeneous evolution behavior (Abidjan from 2011 to 2014 and Cotonou from 2009 to 2014). For both cities and for the temperature indices considered, CDDs are able to explain the large seasonality in consumption data and especially the consumption peak observed in spring and the lower consumption in August. A slightly better relationship is obtained when the heat index is used to calculate CDDs.

Keywords: *Electricity consumption, Cooling degree days, Weather sensitivity, West Africa, Influence.*

Introduction

Electricity Transition in West Africa

Electricity has become a compulsory requirement for enhancing economic activity and improving the quality of human life. Agricultural and industrial production processes are made more efficient through the use of electricity. Households need electricity for many purposes, including cooking, lighting, refrigeration, study and home-based economic activities. Essential facilities, such as hospitals, require electricity for cooling, sterilization, and refrigeration (IRENA, 2012). Electricity is also an important factor of development and a critical component in economic growth (PNUD, 2015). Indeed, the rate of electrification in African countries is lower than in other regions of the world. According to United Nations Development Program (UNDP), it is estimated at 42 per cent; compared to a global rate of 75 per cent (PNUD, 2015). The situation in sub-Saharan Africa is much more precarious with 30 per cent electrical coverage in an urban area versus 14 per cent in rural areas (Pineau, 2008). Furthermore, in West Africa the access to electricity was only 20 per cent in 2012 (DIOP, 2015) with significant differences between the figures of access in urban areas, which are on average 40 per cent, and in rural areas where the figure is between 6 per cent and 8 per cent (ECREEE, 2015). National electricity production is mainly by thermal plants, which will limit the goals set by the states of the region to achieve the electrification by 2030 (REN21, 2014) and (Rojey, 2011).

West Africa's areas have large reserves of fossil fuels such as coal, oil, and natural gas. It is also rich in natural and sustainable energy resources including Variable Renewable Energy (VRE) from solar radiation, water, and wind (ECREEE, 2012). Thus, better electricity support is expected in the next decade. As stated in the last Conferences of Parties (COP 21 in Paris, COP 22, Marrakech), a massive deployment of VRE is actually the only way that can improve electricity access in Africa without negative impact on the climate.

On the other hand, climate change is expected to impact water, wind and solar radiation resources, and hence potential production of VRE. Moreover, warming related to climate change may increase energy demand, but also the risk of system failure, especially as a result of more frequent and intense temperatures (context of heat waves). Energy supply failures can also result from low hydropower production periods (from water reservoirs or river flow) as a result of water resource variability, which is potentially important in many regions worldwide. It is, therefore, imperative to understand the links between climate and energy. In the present work, we explore how the electricity consumption in West African Cities depends on weather variables with the aim to assess the possible climate change effects on the electricity demand.

A commendable number of studies especially in Europe, Asia, and North America, have shown how weather affects electricity demands. For Italy, Scapin *et al.* (2015) could relate the Italian's national electricity demand from 1990 to 2013 to temperature and solar radiation. Their method was at first, to use 3-degree polynomial trend to normalize daily data by removing the non-climatic part of the consumption change. Secondly, they used high-resolution daily temperature, close to cities to be the most representative of the consumption regions. These temperature data were converted in cooling- and heating-degree-days (resp. CDD and HDD). The authors could highlight the strong relationship between normalized consumption and Degree Days (DD). The sensitivity of the consumption to DD was moreover found to present a strong positive trend in the considered period, increasing by a factor 3.5 between 2013 and 1990. This increase is much greater than the increase of the Italian total electricity demand and was related to the growing demand for air conditioning equipment.

On another side, Bessec and Fouquau, (2008) showed in their study, that temperature is a major determinant of electricity consumption in Europe and that the sensitivity of electricity consumption to temperature has increased. The temperature was also found to impact electricity consumption by Hernandez *et al.* (2012) and Bertrand and Reysset (2008). The latter noted a change in France's electricity consumption behavior with a drop of heating due to warming. On the other hand, warming was found to increase needs for air conditioning. Also, Sailor and Muñoz, (1997), developed a methodology for assessing the sensitivity of electricity and natural gas consumption to climate at regional scales. The approach used involved a multiple-regression analysis of historical energy and climate data in the United States.

Isaac and Van Vuuren, (2009) described residential heating and cooling demand for the rst time at a global scale. According to the author's projection, the global energy demand for heating would increase until 2030 when it will stabilize. The writers used relatively simple relationships to define heating and cooling demand and to explore the effects of climate change on this fake energy demand Schipper and Meyers, (1992). According to the others, weather sensitivity increases the energy bills and mitigation challenge. It is decreased in regions with most high-income levels. They also illustrated how heating effects in Africa will increase electricity consumption due to an increased use of air conditioning.

To date, there have been few studies on Africa and even less on West Africa, although it mentioned that the threat on reliability and efficiency of electricity systems was due to more frequent and intense heat waves (Sofia Aivalioti, 2015).

Based on cross-correlation analyses between the electricity consumption, maximum temperature and minimum relative humidity in the air, Adjamagbo *et al.* (2011) highlighted that the dry periods in Togo result in higher electricity consumption.

In the present study, we analyze the dependence of electricity consumption on weather sensitivity for two tropical cities located within the same climatic region of West Africa. Consumption data for the recent years and a suite of temperature indices based on weather data received from the National Meteorological Synoptic Stations are used.

Factors Influencing Electricity Demand

Electricity consumption is classically reported to be first determined by a number of socio-economic and technical factors (Table 10.1). Demography, GDP, and urbanization are often identified as the main socio-economic factors that affected the electricity consumption in western Africa's cities.

Table 10.1: Social and Economic Factors Reported Influencing Electricity Consumption

Factors	Effects on the Demand	Authors
1. Population (Demography)	1,2,3,4 and 5 increase of the demand	(Adom *et al.*, 2012)
2. Gross Domestic Product (GDP)		(Ouédraogo, 2010)
3. Incomes Level	6,7,8 increase or decrease the demand	(Inglesi-Lotz and Blignaut, 2011)
4. Economic growth		
5. Gross Capital Formation		(Akinlo, 2009)
6. Changes in the structure of the economy		(Isaac and Van Vuuren, 2009)
7. Changes in efficiency		
8. Industry efficiency		
9. Degree of urbanization		

Electricity consumption is also reported to depend on climatic factors. A first factor is a day-light time (Hernández *et al.*, 2012). This astronomic factor determines the duration of the day for which lighting is required. It is not expected to have a significance in tropical regions contrary to high latitude regions where large variations in daylight time occur between summer and winter. The seasonal variations of weather are also important drivers of electricity consumption (Akinlo, 2009). In high latitude countries, higher consumption is observed in winter for heating facilities. Higher consumption is also obtained during hot weather sequences for building cooling. In both cases, consumption is highly sensitive to the air temperature. When the current temperature is lower (greater) than a given comfort temperature threshold, the energy demand for heating (cooling) is often considered to be proportional to the difference between the current temperature and the threshold (Isaac and Van Vuuren, 2009). Other weather variables have been reported to potentially influence the demand for heating/cooling. During cold weather, strong wind increases the sensation of cold. During hot weather, humid conditions increase the sensation of heat and the demand for cooling via air conditioning.

Different Energy Demand indices, mainly based on temperature, have been proposed to characterize the weather dependency of the electricity consumption.

The so-called Heating-Degree-Days and Cooling-Degree-Days are often used in the energy sector to estimate the energy demand for respectively building heating/cooling (Adom *et al.*, 2012). A number of different definitions have been proposed (Met Office UK). Classically calculated on a daily basis, they make use of different types of temperature data: mean daily temperature, daily minimum and maximum temperatures, and hourly temperatures (Bessec and Fouquau, 2008).

In the present study, we explore if part of the variations in electricity consumption can be explained by Cooling-Degree-Days (CDD). We consider different estimates of CDD based on raw temperature and different other weather index. We consider especially the Humidex and the Heat Index, both estimated from different combinations of temperature, air humidity, and dew point.

In the first section, we present the consumption and weather data. We next describe the different weather indices used to estimate the CDD. The relationships between CDD and consumption are described and commented in the following section. Section four concludes.

Materials and Methods

Brief Overview of the Study Area and its Selection Criterion

The framework of investigation proposed in this research was conducted on two megacities located on the Western Coast of Africa namely: Abidjan, and Cotonou (Figure 10.1).

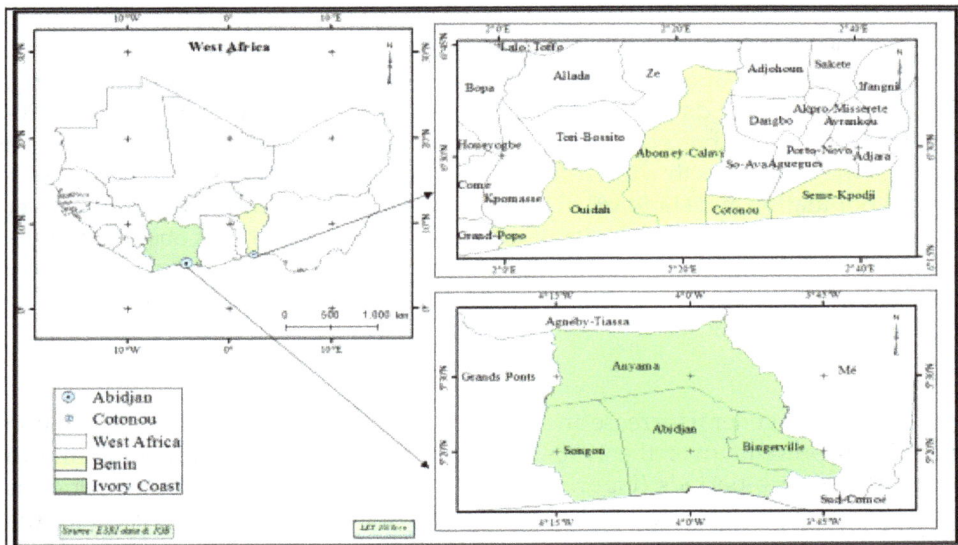

Figure 10.1: Location Map of the Study Area.

The selection of cities was based on the political and economic importance in their countries Côte d'Ivoire (Abidjan) and Cotonou (Benin) (ECA SRO-WA, 2016).

This choice was based on the fact that they have precarious energy coverage and a lack of study based on the climate change service for electricity sector (IEA, 2014 and IEA, 2015).

Abidjan, the economic capital of Ivory Coast has 20 per cent of the country's population (approximately 4.4 million people in 2014). The population of Cotonou was 78000 in 2013. The economic growth rate of both cities is around 2.6 per cent with a rate of urbanization estimated between 3.7 per cent and 4.2 per cent per annum from 2000 to2015, each contributing 3.7 per cent of GDP.

Both cities are situated in Guinean region and especially in the tropical wet climate zone, Cotonou has a tropical savanna climate with dry winters and Abidjan a tropical monsoon climate with short dry season. Its annual rainfall is 1100 to 1600 mm. Relative humidity varies between 32 per cent and 95 per cent. The study area, as most of West African coastal countries, consists on the south by a coastal plain. The wet/dry seasons result from the interaction of two migrating air masses (Ulrike faLK and JörgSzarzynSKi, 2010).

In both cities, annual mean temperature is higher than 26.5 °C and it typically varies from 22°C to 32°C the warm season lasts from January to May. The hottest period of the year is between February and March with mean temperature often higher than 29 °C[1]. From July to September the monthly mean temperature is around 25°C[2]. The Dew point which is often used as a measure of how comfortable a person can feel warm is between 12°C and 28°C.

Electricity Consumption Data

Data on electricity consumption for Abidjan and Cotonou were obtained from National Electrical Companies of each country (SBEE in Benin, CIE in Côte D'Ivoire). Time series monthly data over1990-2015 was received for Abidjan and for 2000-2015for Cotonou. In both countries, the consumption is generally increasing as a result of the GDP growth which also helped the region to have significant growth in both population and standard of life (Adom *et al.*, 2012). The increase is however neither smooth nor progressive and some discontinuities/breaks observed in the general trend. The globally increasing trend is also highly influenced by seasonal variations with higher consumption in the summer and lower consumption in winter for both the locations.

Abidjan

Besides the general increase and seasonal dependence, electricity consumption in Abidjan presents over the 1990-2015 period several phases corresponding to four different sub-periods (Figure 10.2a). The behavior within each sub-period is roughly homogeneous (stationary or progressive increase), but the consumption regime clearly changes from one sub-period to the other, sometimes abruptly. More precisely, the first period from 1990 to 1995 is roughly stationary with rather constant

---

1 https://weatherspark.com entered on December 3, 2016 at 10 AM.

2 http://www.weatheronline.co.uk/Africa.htm entered on December 3, 2016 at 11 AM.

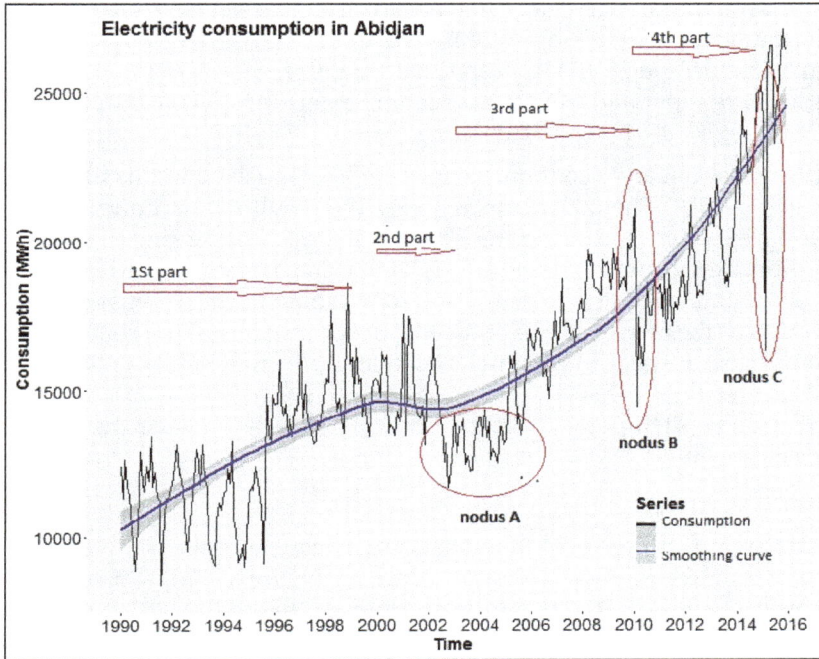

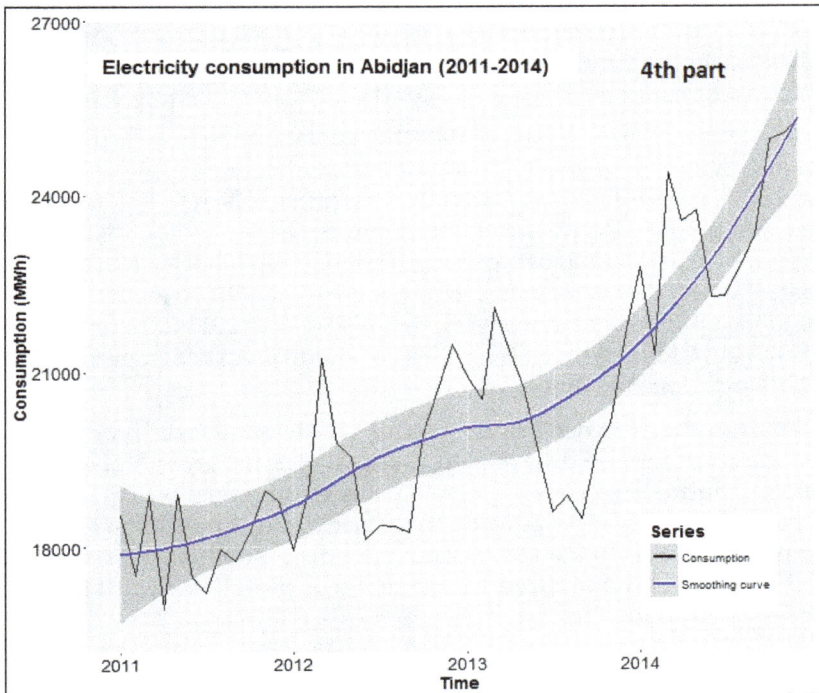

Figure 10.2: Time Series of Monthly Electricity Consumption for Abidjan (A: 1990 to 2015) and Working Part Period (B: 2011 to 2014).

consumption at the annual scale. After an abrupt increase in 1995, the 2nd period, still roughly stationary, extends up to 2002. The third and fourth phases are observed through2003-2010 and 2010-2015 respectively with a fast consumption growth. Both sub-periods are separated from the preceding by an abrupt consumption decrease (2003, 2010).

Different social and economic factors are likely to explain this time evolution (Ouédraogo, 2010). The 2003 drop is likely the result of the political instability the country experienced during the 2002-2004 period. According to the Technical Distribution Department (TDD) of CIE, the period was characterized by several load sheds with blackouts in electricity distribution. The 2010 demand drop has to be related to (*i*) the failure of one of the main power plants of the country resulting in a long period of shedding, energy deficit and (*ii*) a reduction of socio-economic activity as a result of the socio-economic crisis that the country faced at that time. The 2015 drop was temporary, due to technical problems and also the high amount of the electricity bill.

Cotonou

2000 to the 2015-time evolution of electricity consumption in Cotonou is much more smooth and progressive than in Abidjan (Figure 10.3a). It presents again a significantly increasing trend with a significant seasonality (higher demand in winter). Although not as distinctas for Abidjan, 3 different phases could be identified: two main sub-periods with a continuous and homogeneous increase (from 2000 to 2005; from 2008 to 2015) separated by a transition period (2006-2007) with a slightly reduced increasing rate and a lower seasonal variability. No clear socio-economic or technical reasons could be found so far for the 2006-2008 suspected discontinuity.

For both cities, the evolution of monthly consumption is at first governed by non-climatic factors. Our objective is to analyze the climate dependence of the consumption. To be not contaminated by non-climatic factors in the analysis, we extracted sub-periods fromthe previous time series for which the evolution in consumption was estimated to be pseudo-stationary. In other words, an increasing trend may exist but the increase has to be progressive with no abrupt change from one year to the other. The retained work periods are then 2011-2014 for Abidjan and from 2009 to 2014 for Cotonou. They thus cover 4 to 6 years of a period estimated to be relatively homogeneous.

To analyze the weather driven part of the demand, we have additionally removed the consumption drift related to global product growth of each country. We, therefore,normalized consumption data. A trend model was fitted to the raw consumption data over the working period. The trend was then removed as follows: each monthly data was divided by the corresponding monthly value obtained from the trend model. The normalized (detrended) consumption data (Figure 10.3) is thus expressed as following.

$$N\_C(t) = C(t)/T(t) \tag{1}$$

where,

$T(t)$ is the trend value of the monthly consumption for month t.

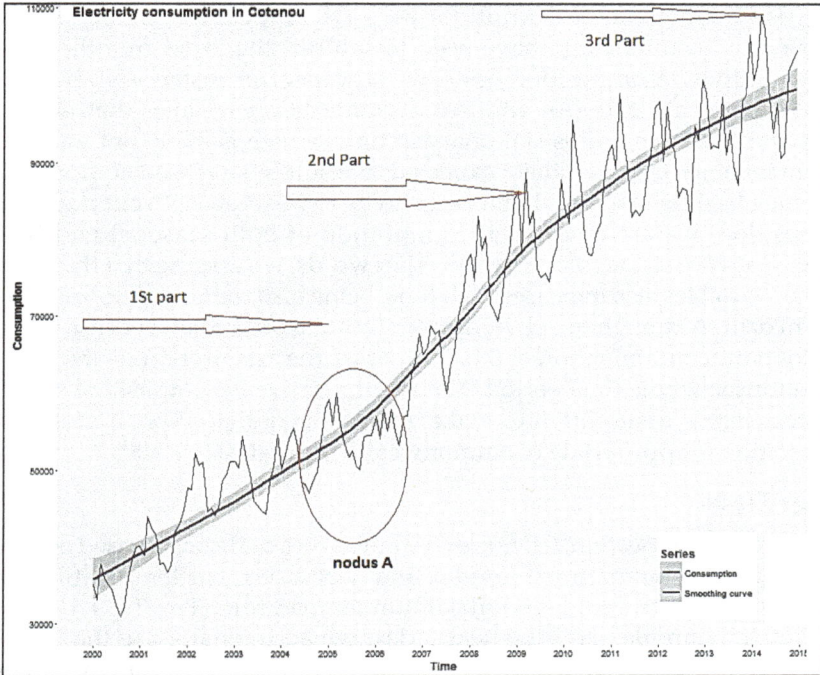

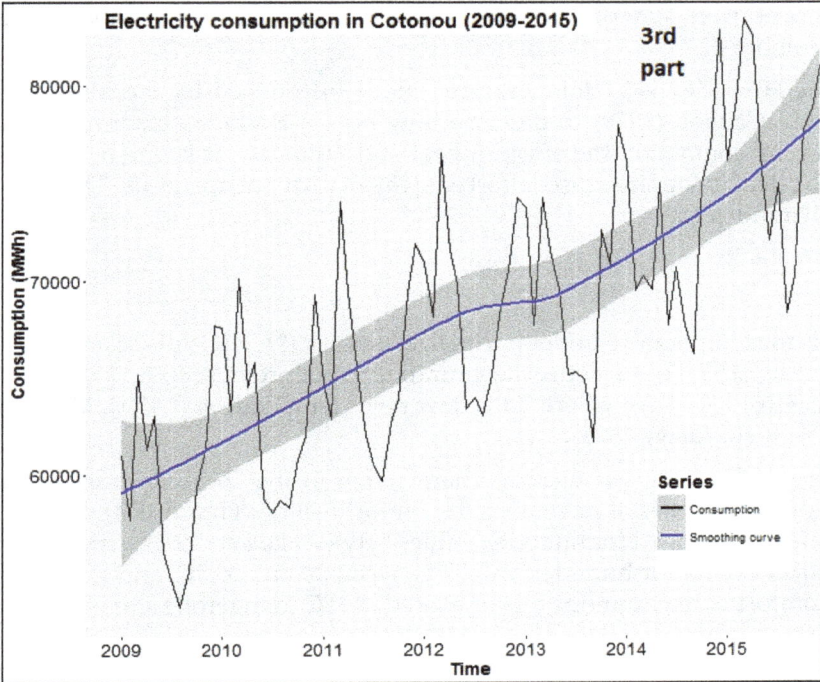

Figure 10.3: Cotonou Electricity Consumption Time Series (A: 1990 to 2015) and Working Part Period (B: 2011 to 2014).

We use there a classical Multiplicative Decomposition of the time series. Another normalization could have been to subtract the trend fromthe raw data. Such an additive decomposition has been proposed for instance by Bessec *et al.*, 2008 or Scapin *et al.*, 2016. An additive decomposition assumes that the climate-driven part of the demand is stationary in time, which is likely not the case. The growth in incomes level is indeed expected to result in increased equipment in air conditioning facilities and thus higher sensitivity to weather. With our multiplicative decomposition, we assume that the amplitude of both seasonal and irregular variations increase as the trend rises. In other words, we assume that the sensitivity to weather variables also increases with time, being also roughly proportional to the long-term drift. Although not obvious from the data of Abidjan because of the high socio-economic contamination of the time series, the assumption seems to hold for Cotonou for the whole time period. The multiplicative decomposition seems thus to be a reasonable assumption to make. In such a configuration, a multiplicative decomposition is appropriate (Commonwealth of Australia, 2008[3]).

Weather Data

In the following section, CDD are computed on a daily basis based on different weather indices: temperature, humidex and heat index. For temperature, we use either daily mean, minimum and maximum temperature (Tm, Tx, Tn (Robinson, 2001)). For the humidex and the heat index, we additionally use the Daily Dew Point Temperature (Td) and the Daily Relative Humidity (R). Daily weather data are obtained from synoptic stations. They cover the 1980-2014 period for Abidjan and Cotonou.

The Humidex is a "temperature" index introduced by the Meteorological Service of Canada (1979) to describe how hot an average person perceives the weather, by combining the effect of heat and humidity. It is one of the common ways to calculate summer discomfort equivalent to dry temperature. The expression of the Humidex is:

$$H = T + R \tag{2}$$

where,

T is the temperature under normalized shelter ($^\circ$ C); R is the relative humidity such that $h = 0.55 * (e-10)$, with e the saturating vapor pressure $e = 6.11 \times expr (5417.75 \times ((1/273.16) - (1/Td)))$ where Td is dew point temperature ($^\circ$ C) (J. M. Masterton and F. A. Richardson, 1979).

This index has been used to characterize in the (Temperature, Humidity) space, different zones of increasing discomfort feeling (Figure 10.4). A low risk of discomfort is obtained for humidex values between 20 and 29°C; a medium risk of discomfort occurs for humidex values between 30 and 39 °C; and maximum risk of discomfort when humidex hits or exceeds 46 °C (dangerous situation) (Rome *et al.*, 2016)).

---

3 Commonwealth of Australia, 2008: http://www.abs.gov.au/websitedbs/D3310114.nsf/home/ Time+Series+Analysis:+The+Basics.

Table 10.2: Degrees of Comfort Based on Humidex indexes

Indice HUMIDEX

		Température (°C)															
		21	25	30	31	32	33	34	35	36	37	38	39	40	41	42	43
Humidité relative (%)	20											40	41	43	44	46	47
	30			31	33	34	36	37	38	40	42	43	45	47	48	50	51
	40		26	34	35	37	39	40	42	44	45	47	49	51	53	54	56
	50	22	28	36	38	40	41	43	45	47	49	51	53	55	57		
	60	24	30	38	40	42	44	46	48	50	52	54	57				
	70	25	32	41	43	45	47	49	51	53	56	58					
	80	26	33	43	45	47	50	52	54	57	59						
	90	28	35	45	48	50	52	55	57	60							
	100	29	37	48	50	53	55	58									

The Heat Index, introduced by George Winterling in 1978 has a similar definition to that of the Humidex. It can also be defined as the apparent temperature, corresponding again to the temperature felt by the human body when relative humidity is combined with the air temperature.[4] (Oueslati *et al.*, 2016). Its expression is the following:

Heat Index (HI) = -42.37 + 2.05 T + 10.14 R - 0.22 T R - 6.83 $10^{-3}T^2$ - 5.48 $10^{-2}R^2$ + 1.23 $10^{-3}T^2R$ + 8.53 $10^{-4}T$ R^2 - 1.99 $10^{-6}T^2R^2$

with T the air temperature (°F) and R - relative humidity (per cent) (Robinson, 2001).   (3)

Cooling Degree Days

Heating degree days and Cooling Degree Days are used extensively in the energy sector for estimating the energy consumption from buildings. (http://www.degreedays.net/introduction Martin Bromley, 2009)

When HDDfocuses on heating requirements during cold weather, CDDfocuses on cooling requirements during hot weather. CDDare a measure of how much (in degrees), and for how long (in days), the outside air temperature was above a certain level[5]. They estimate, for a given period of time, the cumulative exceedance of temperature above (for CDD) or below (for HDD) a given temperature threshold (Aceituno, 1979)[6].

---

4 (US Department of Commerce, 2016)https://www.weather.gov/ama/heatindex.

5 http://www.degreedays.net/introduction#Other_types_of_degree_days.

6 http://www.degreedays.net/regression-analysis.

CDD can be estimated on a daily basis. Different formulations have been proposed based on different temperature data. When estimated from the mean daily temperature Tm, the CCD are simply obtained as:

CDD(Tm) = (Tm – Tr) if Tr < Tm

CDD(Tm) = 0 if Tm < Tr (4)

Tr is the temperature threshold, classically taken as 25°C.

When hourly temperature data are available, the same formulation is used for each hour of the day and the CDD for the day are just obtained as the mean of those hourly estimates. The last formulation is based on the minimum (Tn) and maximum (Tx) daily air temperatures which are also often available (Aceituno, 1979) and (Santee and Wallace, 2005).

CDD(Tx,Tn) = ((Tx + Tn)/2) – Tr) if Tr < Tn

CDD(Tx,Tn) = (Tx - Tr) * (0.08 + 0.42 * (Tx - Tr)/(Tx - Tn))
**   if Tn < Tr < Tx, CDD(Tx,Tn) = 0 if Tx <Tr** (5)

In the following, we will consider 6 different estimations of the CDD based on the different available CDD formulations and the different temperature indexes: air temperature, humidex, heat index. The three first CDD estimations follow (4). They use as temperature index the value obtained with the mean daily value of temperature. In this configuration, the daily value of the humidex and the heat index are obtained from the daily value of air temperature, relative humidity and dew point temperature. The estimation obtained with air temperature, humidex and heat index will be respectively referred to as CDDTm, CDDHUm, CDDHIm.

The three other estimations make use of the minimum and maximum temperature estimates available each day. Thus they follow (5). For this estimation, the Humidex is estimated for both the minimum and maximum temperatures of each day. For each day, these two Humidex values are used as temperature indices in (5) for Tx and Tn respectively. The same procedure is followed for the Heat Index. The Heat Index is estimated for both the minimum and maximum temperatures of the day. These 2 estimates are used for the CDD calculation in (5). The CDD estimates obtained respectively with minimum and maximum temperatures, humidex, and heat index arereferred to as CDDTxn, CDDHUxn, CDDHIxn.

Results and Discussion

The time series data of normalized monthly consumption are displayed in Figure 10.4. They highlight the seasonality of the demand. For both cities, there is a peak in every March, and a trough in every August. The seasonal fluctuations are roughly constant in size and comfort as per the assumption of multiplicative decomposition proposed previously.

CDD were estimated at adaily time step. The daily time series data off CDD estimates were aggregated to monthly time step. The monthly aggregated CDD were next compared to the monthly normalized consumption.

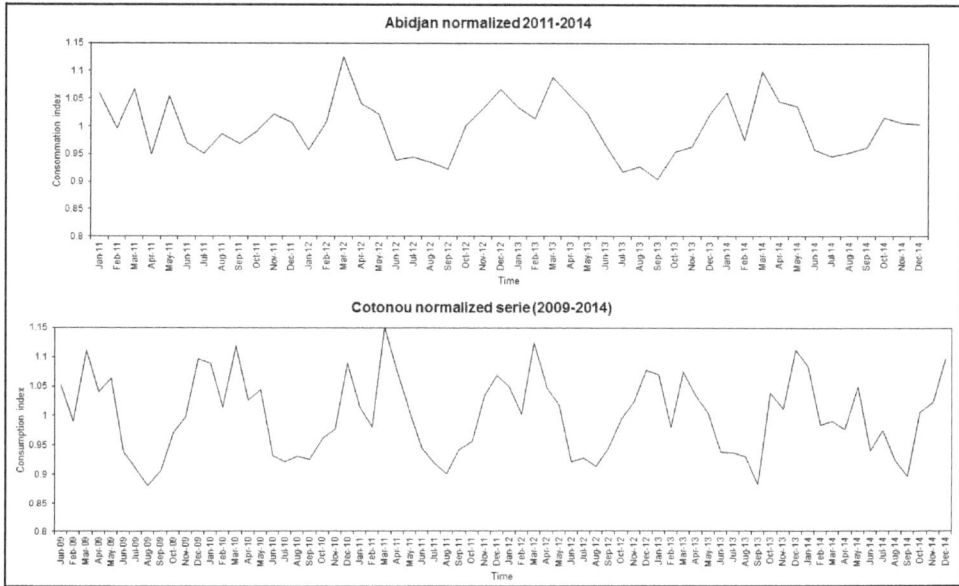

Figure 10.4: Normalized Time Series Data of Electricity Consumption of Abidjan and Cotonou.

Figures 10.5a-b presents for both cities the inter- annual evolution of monthly CDDs and monthly electricity consumption.

There is in cases, an increase of the consumption and the temperature indices from January with a peak in March (highest point) and a decreasing trend from May to August (lowest point). Whatever be the index, the profile of seasonal variations of consumption is roughly follow those of CDDs. Temperature and electricity consumption, therefore, appear to vary together.

This is also highlighted in the scatter plots in Figure 10.6a-b which shows the monthly consumption as a function of the different monthly estimated CCD.

To better quantify the relationships, the Pearson correlation coefficient (r) was calculated and presented in Table 10.3.

Table 10.3: Pearson Correlation Test (P value < 0.001) for the Electricity Demand and Weather Sensitivity in Abidjan and Cotonou

Country		Correlation Coefficient					
		CDDTm	CDDHUm	CDDHlm	CDDTxn	CDDHUxn	CDDHlxn
Abidjan	Conso index	0.78***	0.73***	0.76***	0.78***	0.73***	0.76***
Cotonou		0.76***	0.75***	0.81***	0.76***	0.75***	0.84***

*** Indicate a significant higher correlation at 0.1 per cent.

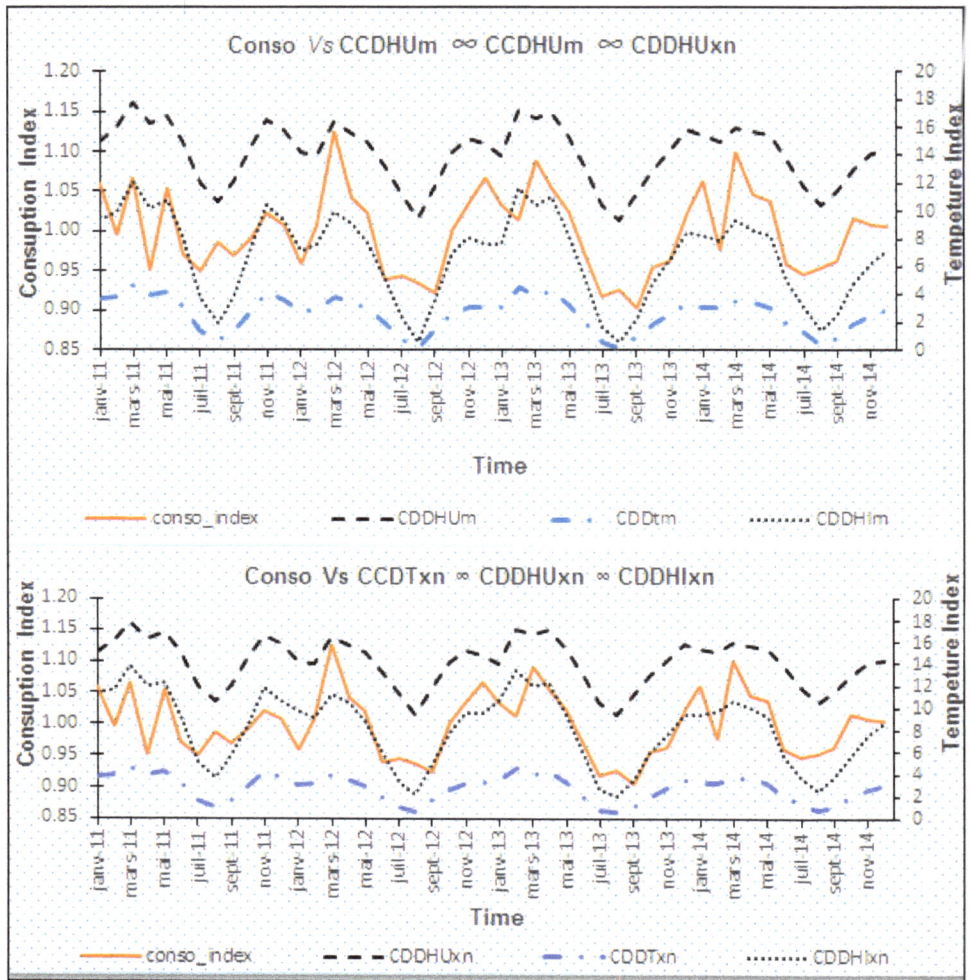

Figure 10.5a: Inter-annual Evolution of Monthly CDDs and Monthly Electricity Consumption in Abidjan.

In all cases the correlation coefficient is higher than 0.76 and is highly significant with a P-value lower than 0.001. For Abidjan, the best correlations are obtained with temperature data only. No difference is obtained between the two CDD formulations: the min/max temperatures do not give additional information when compared to the mean daily temperature. For Cotonou, the best correlations are obtained with the Heat Index. They are significantly higher than those obtained with the temperature alone, highlighting the role of humidity in the consumption. They are also always significantly higher than the correlation with the Humidex,

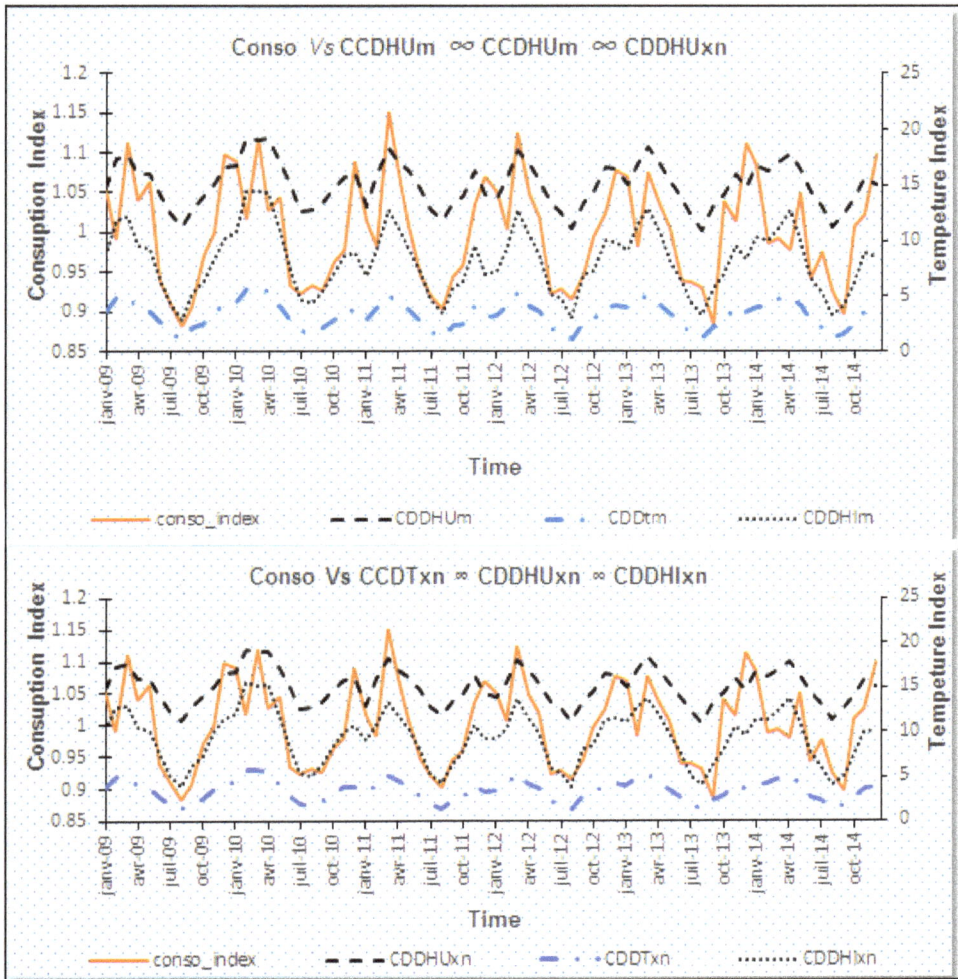

Figure 10.5b: Inter-annual Evolution of Monthly CDDs and Monthly Electricity Consumption in Cotonou.

highlighting the better predictive power of Heat Index for this area. Using the min and max temperature to derive respectively min and max Heat Indices for each day is also found to produce better correlation.

Electricity consumption rate is therefore also linked to weather conditions in these two African cities. Hot conditions lead to increased consumption. This likely has to be related to the use of electric cooling materials to improve the conformability of the population (fan for the mean standing and air condition for high standing). In Cotonou, the perceived temperature also seems to increase with relative humidity.

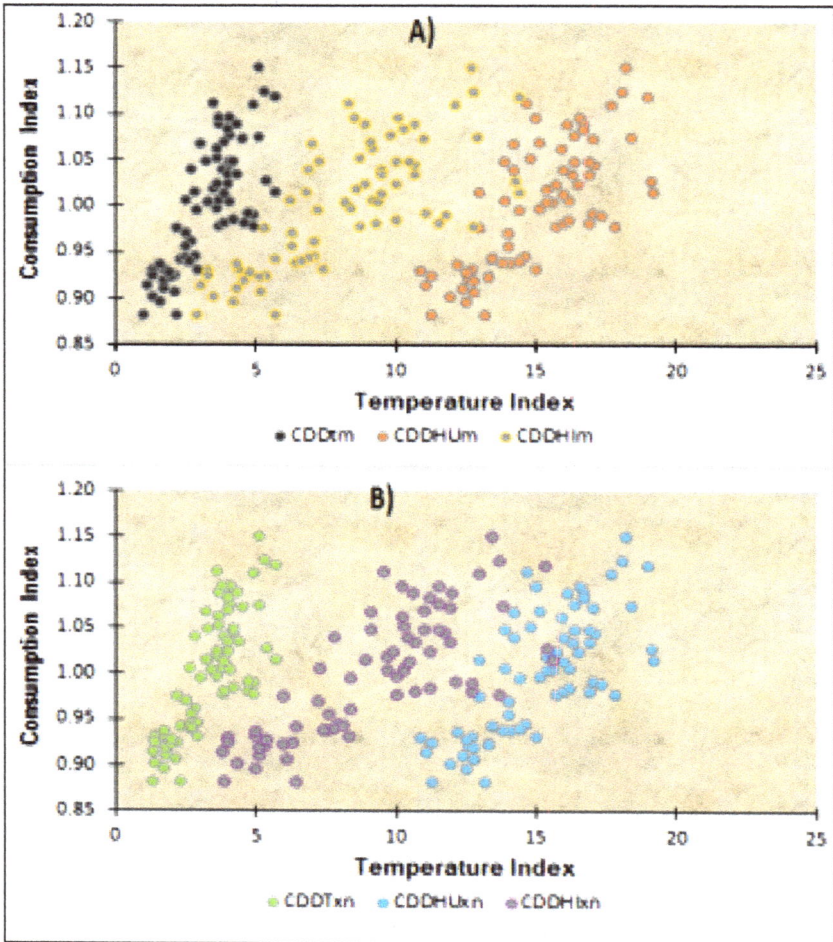

Figure 10.6a: Scatter Plot of Consumption against the Temperature Index Series from 2011 to 2014 in Abidjan.

Conclusions and Perspectives

Many factors including GDP, urbanization, population growth affect the electricity consumption. This article is among the first establishing the relation between electricity consumption in West African Cities and weather sensitivity in a view to assessing the potential impact of climate change on the electricity demand in the future. Results showed that the seasonal fluctuations of the monthly normalized consumption from 2009 to 2014 in Cotonou and from 2011 to 2014 in Abidjan are unevenly constant in size. This revealed that for both cities, electricity increases from January to March, while it decreases from July to September. The annual time series data of temperature index and electricity consumption displays that the

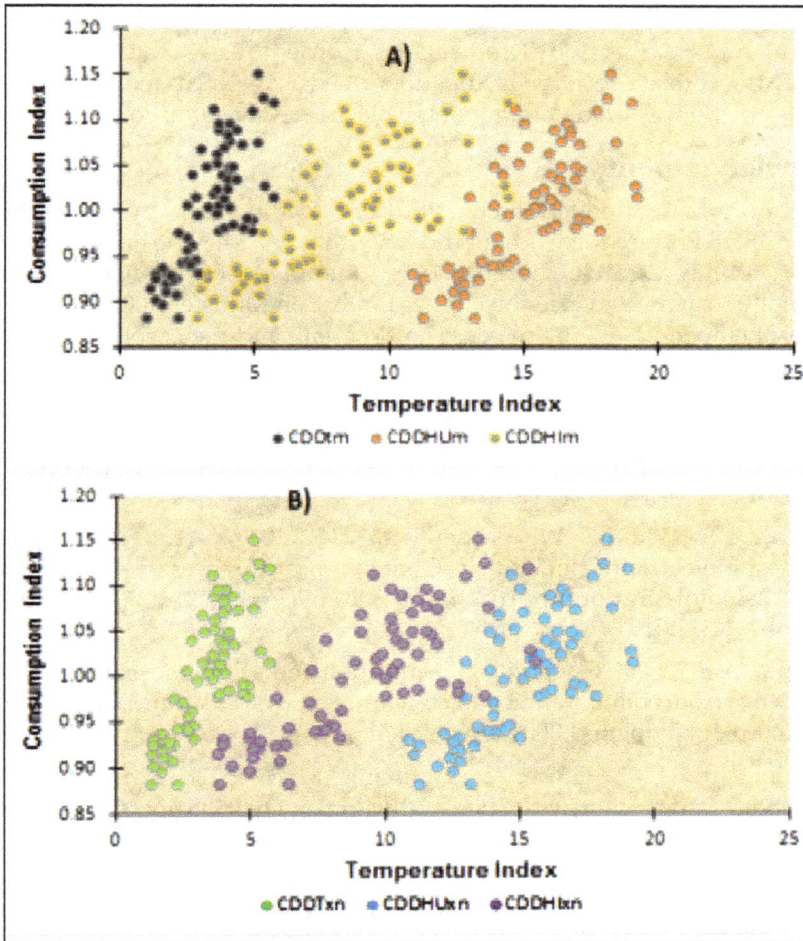

Figure 10.6b: Scatter Plot of Consumption against the Temperature Index Series from 2009 to 2014 in Cotonou.

temperature and electricity consumption fluctuate in the same way. Consequently, temperature and humidity (CDDTm, CDDHIm, CDDTxn, CDDHIxn, CDDHUm and CDDHUxn) are linked to the electricity consumption. The temperature index with Humidex seems less correlated to the electricity consumption than the others. This means that the combination of temperature and relative humidity affects the electricity consumption more than the combination of temperature and dew point.

Based on these results, a further investigation taking into account other potential predictors should be implemented. A better use, if possible of the whole consumption time series could improve climatic signal detection for electricity demand. In addition, the assessment of the sensitivity of the result to the temperature threshold will be needed to qualify its effects on the consumption. It will be obviously also useful to extend this study to other cities in order to assess the regional stability of the relationships. The relationships identified from such analyses could lead finally to a first guess prediction model of electricity consumption in this area

Acknowledgements

The authors acknowledge contributions to the study by the Director of CCBAD/UFHB Postgraduate School and CRNI/LAPA_MF who hosted the research laboratory for this research. The authors are grateful to the National Electricity and Meteorology services and also the Higher Education Minister of Côte d'Ivoire and Benin for the open access to the consumption data and meteorological data used in the present work.

References

1. Aceituno, P. (1979). Statistical formula to estimate heating or cooling degree-days. *Agric. Meteorol.* 20, 227–232

2. Adom, P.K., Bekoe, W., Akoena, S.K.K. (2012). Modelling aggregate domestic electricity demand in Ghana: An autoregressive distributed lag bounds cointegration approach. *Energy Policy* 42, 530–537. doi: 10.1016/j. enpol.2011.12.019

3. Akinlo, A.E. (2009). Electricity consumption and economic growth in Nigeria: Evidence from cointegration and co-feature analysis. *Research Gate* 31, 681–693. doi: 10.1016/j.jpolmod.2009.03.004

4. Babonneau, F.L.F., Haurie, A., Caramanis, M. (2015). ETEM-SG: Optimizing Regional Smart Energy System with Power Distribution Constraints and Options. Springer Verlag.

5. Bessec, M., Fouquau, J. (2008). The non-linear link between electricity consumption and temperature in Europe: A threshold panel approach. *Energy Econ.* 30, 2705–2721. doi: 10.1016/j.eneco.2008.02.003

6. Gupta, E. (2016). The effect of development on the climate sensitivity of electricity demand in India. *Clim. Change Econ.* 7, 1650003. doi:10.1142/ S2010007816500032

7. Chen, K., Bi, J., Chen, J., Chen, X., Huang, L., Zhou, L. (2015). Influence of heat wave definitions to the added effect of heat waves on daily mortality in Nanjing, China. *Sci. Total Environ.* 506–507, 18–25. doi: 10.1016/j.scitotenv.2014.10.092

8. Christenson, M., Manz, H., Gyalistras, D. (2006). Climate warming impact on degree-days and building energy demand in Switzerland. *Energy Convers. Manag.* 47, 671–686.

9. Fu, K.S., Allen, M.R., Archibald, R.K. (2015). Evaluating the Relationship between the Population Trends, Prices, Heat Waves, and the Demands of Energy Consumption in Cities. *Sustainability* 7, 15284–15301. doi:10.3390/su71115284

10. Hernández, L., Baladrón, C., Aguiar, J.M., Calavia, L., Carro, B., Sánchez-Esguevillas, A., Cook, D.J., Chinarro, D., Gómez, J. (2012). A Study of the Relationship between Weather Variables and Electric Power Demand inside a Smart Grid/Smart World Framework. Sensors 12, 11571–11591. doi:10.3390/s120911571

11. Higley, L.G., Pedigo, L.P., Ostlie, K.R. (1986). DEGDAY: a program for calculating degree-days, and assumptions behind the degree-day approach. *Environ. Entomol.* 15, 999–1016.

12. Holland, R.A., Scott, K.A., Floerke, M., Brown, G., Ewers, R.M., Farmer, E., Kapos, V., Muggeridge, A., Scharlemann, J.P.W., Taylor, G., Barrett, J., Eigenbrod, F. (2015). Global impacts of energy demand on the freshwater resources of nations. *Proc. Natl. Acad. Sci. U. S. A.* 112, E6707–E6716. doi:10.1073/pnas.1507701112

13. IRENA (2012). Prospects_for_the_African_PowerSector.pdf [WWW Document]. URL https://www.irena.org/DocumentDownloads/Publications/Prospects_for_the_African_PowerSector.pdf (accessed 10.31.16)

14. Isaac, M., Van Vuuren, D.P. (2009). Modeling global residential sector energy demand for heating and air conditioning in the context of climate change. *Energy Policy* 37, 507–521.

15. Jiang, F., Li, X., Wei, B., Hu, R., Li, Z. (2009). Observed trends of heating and cooling degree-days in Xinjiang Province, China. *Theor. Appl. Climatol.* 97, 349–360.

16. J. M. Masterton, F. A. Richardson (1979). Humidex, A Method of Quantifying Human Discomfort Due to Excessive Heat and Humidity, Downsview, Ontario, Environnement Canada 45.

17. Ouédraogo, I.M. (2010). Electricity consumption and economic growth in Burkina Faso: A cointegration analysis. *ResearchGate* 32, 524–531. doi: 10.1016/j.eneco.2009.08.011

18. Oueslati, B., Sambou, M.-J.G., Pohl, B., Rome, S., Moron, V., Janicot, S. (2016). Les vagues de chaleur au Sahel/: caractérisation, mécanismes, prévisibilité. Presented at the 29ème Colloque de l'Association Internationale de Climatologie, Association Internationale de Climatologie, pp. 327–332.

19. PNUD (2015). Année d'Action Mondiale (Annuel). Bureau des relations extérieures et du plaidoyer Programme des Nations Unies pour le développement, New York.

20. RINGARD, J. (2013). Etude rétrospective et prospective des vagues de chaleur en Afrique de l'Ouest.

21. Ringard, J., Dieppois, B., Rome, S., Diedhiou, A., Pellarin, T., Konaré, A., Diawara, A., Konaté, D., Dje, B.K., Katiellou, G.L., others (2016). The intensification of thermal extremes in west Africa. *Glob. Planet. Change* 139, 66–77.

22. Ringard, J., Dieppois, B., Rome, S., Dje kouakou, B., Konate, D., Katiellou, G.L., Lazoumar, R.H., Bousou-moussa, I., Konare, A., Diawara, A., Ochou, A.D., Assamoi, P., Camara, M., Diongue, A., Descroix, L., Diedhiou, A. (2014). Evolution des pics de températures en Afrique de l'ouest : étude comparative entre Abidjan et Niamey [www document]. url http://www.academia. edu/15904603/per cent c3 per cent 89volution_des_pics_de_temp per cent c3 per cent 89ratures_en_afrique_de_l_ouest_ per cent c3 per cent 89tude_ comparative_entre_abidjan_et_niamey (accessed 7.22.16).

23. Robinson, P.J. (2001). On the Definition of a Heat Wave. J. Appl. Meteorol. 40, 762–775. doi:10.1175/1520-0450(2001)040<0762: OTDOAH>2.0.CO;2

24. Rome, S., Oueslati, B., Moron, V., Pohl, B., Diedhiou, A. (2016). Les vagues de chaleur au Sahel: définition et principales caractéristiques spatio-temporelles (1973-2014)., in: 29ème Colloque de l'Association Internationale de Climatologie. Association Internationale de Climatologie, pp. 345–350.

25. Santee, W.R., Wallace, R.F. (2005). Comparison of weather service heat indices using a thermal model. *Research Gate* 30, 65–72. doi: 10.1016/j.jtherbio.2004.07.003

26. Scapin, S., Apadula, F., Brunetti, M., Maugeri, M. (2015). High-resolution temperature fields to evaluate the response of italian electricity demand to meteorological variables: an example of climate service for the energy sector. *Theor. Appl. Climatol.* 1–14.

27. SEFA (2015). Sustainable Energy Fund for Africa (SEFA): Annual report 2015.

28. Shumway, R.H. (1988). Applied Time Series Analysis. Prentice-Hall, Englewood Cliffs, NJ.

29. Shumway, R.H., Stoffer, D.S. (2010). Time series analysis and its applications: with R examples. Springer Science and Business Media.

30. Tavish Srivastava (2015). A Complete Tutorial on Time Series Modeling in R. Anal. Vidhya.

31. US Department of Commerce, N. (2016). What is the heat index? [WWW Document]. URL https://www.weather.gov/ama/heatindex (accessed 11.18.16).

32. Van der Zwaan, B., Kober, T., Calderon, S., Clarke, L., Daenzer, K., Kitous, A., Labriet, M., Lucena, A.F., Octaviano, C., Di Sbroiavacca, N. (2016). Energy technology roll-out for climate change mitigation: a multi-model study for Latin America. *Energy Econ.* 56, 526–542.

33. WEO (2015). Africa Could Lead World on Green Energy [WWW Document]. URL http://www.climatecentral.org/news/africa-could-lead-world-on-green-energy-19732 (accessed 11.17.16).

34. Yousefi, M., Damghani, A.M., Khoramivafa, M. (2016). Comparison greenhouse gas (GHG) emissions and global warming potential (GWP) effect of energy use in different wheat agroecosystems in Iran. *Environ. Sci. Pollut. Res.* 23, 7390–7397. doi:10.1007/s11356-015-5964

Chapter 11

Climate Change: Challenges and Mitigation Opportunities in Telecommunications Sector of Emerging Economies

C.S. Azad

National Executive Council Member, Society of Energy Engineers and Managers (SEEM),
Asst. General Manager, Bharat Sanchar Nigam Ltd.
(Govt. of India Enterprise), India
E-mail: csazad68@gmail.com

Abstract

The glorious growth in telecom sector has created many adverse effects on environment like emission of greenhouse gases and carbon dioxide into atmosphere. This growth however has and continues to be at the cost of climate powered by an unsustainable and inefficient model of energy usages. With increasing pervasiveness of mobile phones and the widespread adoption of Information and Communications Technology (ICT) worldwide, the ICT sector is expected to contribute around 3 per cent of the global emissions of Greenhouse Gases (GHG) by the year 2020. The telecommunication sector can help in climate change mitigation by reducing the sector's energy requirements, using renewable energy sources. This paper identifies the climate change challenges and opportunities for mitigation in telecommunication sector with a business model for alternative energy to telecommunication infrastructure in emerging economies.

Keywords: *Climate change, Greenhouse Gas (GHG), Information and Communications Technology (ICT), Mitigation, Alternative energy, Green telecom network, Renewable Energy Service Providing Companies (RESCO).*

Introduction

Telecommunication has become an integral part of our lives. It is difficult to imagine the life without phones now. The demand for telephones keeps growing, and it is coming from the rural and remote areas where access to electric grids is not always guaranteed. Wherever available, grid power is of poor quality and availability. Moreover, the telecom service providers are under tremendous pressure to deliver the solutions in an efficient manner and somehow make network available. Not only does it have to be available, it must be up and running round the clock because it is assumed to be an essential service. No other service, be it healthcare or water supply, is available 24 hours a day. To overcome the electric grid supply problems, the widely deployed solution is to provide diesel generators to power the telecom infrastructure. It also has the disadvantage of increasing carbon emission, which has a negative impact on the environment.

Impact of the Telecom Sector on Climate Change

The primary carbon footprint, or direct impact for a telecom service provider, would include network operational cost, building lighting, and cooling or heating and transportation. The service provider generally has direct control over these. The secondary footprint, a measure of the indirect CO_2 emissions, is associated with the manufacture and eventual breakdown during the whole life cycle of the products that are used. Energy consumed in the manufacture of equipment—for instance, by a Base Transreceiver Station (BTS)—is a source of indirect emission for service providers that use it. Based on International Telecommunication Union (2007) estimation, the Information and Communication Technology (ICT) contributes 2-2.5 per cent into the worldwide greenhouse gas emissions.

Kumar R. and Mieritz, L.(2007) estimated that ICT equipment (excluding broadcasting) contributed around 2 per cent to 2.5 per cent of worldwide Greenhouse Gas (GHG) emissions – 40 per cent of this was reported to be due to the energy requirements of PCs and monitors, 23 per cent to data centres, 24 per cent to fixed and mobile telecommunications, and 6 per cent to printers. Malmodin (2009) used life cycle assessment and results were broadly similar.

Sutherland (2011) quoted the estimates of the overall contribution of the ICT sector to global emission. The estimates are as follows:

1. Gartner Group 2007: ICT sector - 2.5 per cent of global emissions;
2. KTH Royal Institute of Technology 2007: ICT sector - 1.3 per cent of global emissions (3.9 per cent of global electricity use), Entertainment and Media (E and M) - 1.7 per cent of global emissions (3.2 per cent of global electricity use);
3. MIC Japan 2006: ICT sector - 2.2 per cent of national emissions;
4. Australian Computer Society 2010: ICT sector - 2.7 per cent of national emissions (> 7 per cent of electricity used)

Gerhard Fettweis and Ernesto Zimmermann (2008) estimated that currently, server farms and telecommunications Infrastructure (Fixed and mobile) are responsible for roughly 3 per cent of the world wide electricity consumption. If the present growth trend of 16 per cent per year continues, as the increase of internet traffic and the number of mobile phone subscribers suggests, this consumption rises by a factor of 30 in only 23 years. Further, the energy consumption of base stations and backhaul networks of cellular communication networks was referred as approx. 60 billion kWh per year corresponding to 0.33 per cent of global electricity consumption. Figure 11.1 explains the components of ICT and their CO_2 emission.

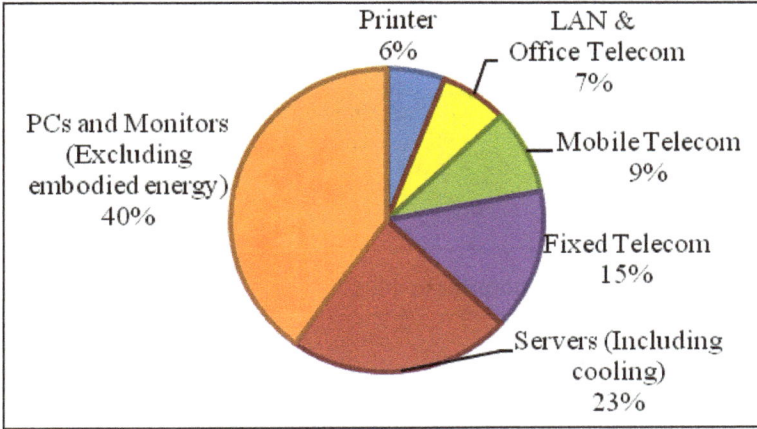

Figure 11.1: Components of ICT and their CO_2 Emission.

The access networks fixed and mobile radio consume largest shares of the overall energy consumption of telecommunication networks owing to lots of active network elements that are widely distributed throughout the field. Lange C *et al.* (2011)

Telecom networks and services are expanding at an exponential scale all over the world. Figure 11.2 demonstrates the projection of CO_2 footprint due to various telecom services in year 2020.

The various greenhouse gases emitted in telecommunication sector with their global warming potential have been depicted in Table 11.1.

Table 11.1: Greenhouse Gases in Telecommunications

Name	Equivalence to Carbon Dioxide	Source
Hydrofluorocarbons (HFCs)	11700	Refrigerants, propellants and cleaners
Sulphur Hexafluoride (SF$_6$)	23900	Electrical insulation
Perfluorocarbons (PFCs)	6500	Refrigerants and fire suppression systems
Nitrous Oxide (N$_2$O)	310	Vehicle engines and power generation
Methane (CH$_4$)	21	-
Carbon Dioxide (CO$_2$)	1	Vehicle engines and power generation

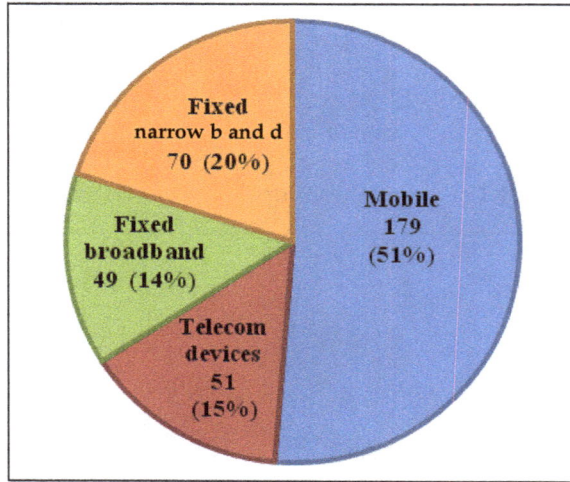

Figure 11.2: Global Telecom Footprint in 2020.

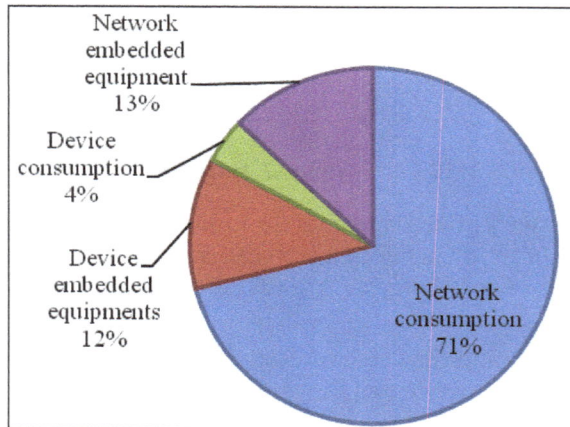

Figure 11.3: Energy Consumption Distribution in Telecom Sector.

Key Challenges for Telecon Infrastructure in Developing Countries/Emerging Economies

1. Lack of energy infrastructure in rural areas: Deficient grid power makes it imperative to use non-grid sources, the most common being diesel generators.

2. Lack of effective incentives for alternative energy sources: There is a general feeling that the alternate sources of energy are more expensive compared to the grid electricity or even that obtained through diesel generators. Incentives are required for using Renewable Energy (RE) and energy efficiency measures for telecom infrastructure.

3. Lack of incentives for Waste Minimisation: Incentive is needed to promote waste minimisation and recycling in the telecom industry.

Mitigation Opportunites for Telecom Sector

* **Use of alternative energy:** The move from diesel to solar and other alternate sources of energy will result in a reduction of CO_2 emissions as well as savings in operating expenses for telecom tower companies. Moving to alternative energy sources can generate millions of carbon credits that could offset the operating expenses on their towers. In addition, savings in the energy bills would further reduce the operating expense.

* **Green telecom networks:** In telecom networks, greening refers to minimising the consumption of energy through the use of energy-efficient technology, using RE sources and eco-friendly consumables.

* **Green manufacturing:** The greening process involves using eco-friendly components, energy-efficient manufacturing equipment, electronic and mechanical waste recycling and disposal, reduction in use of hazardous substances like chromium, lead and mercury and reduction of harmful radio emission.

* **Design of green central office buildings:** Optimisation of energy power consumption and thermal emission, minimisation of GHG emission.

* **Waste disposal:** Disposal of mobile phones, network equipment, *etc.*, in an environment-friendly manner so that any toxic material used during production does not get channelized into the atmosphere or undergroundwater.

Business Model for Green Power to Telecom Towers in Rural Areas of Developing Countries

It is proposed that Government must provide subsidy to village cooperatives or companies interested in providing green power in villages. This Renewable Energy Service Providing Companies (RESCO) must be mandated to prioritize and expedite supply of clean energy for telecom towers. There must be specific roll-out commitments on RESCOs to provide energy to telecom infrastructure companies at agreed prices. With telecom services available in several areas where there is no electricity, it makes eminent sense to require RESCOs to prioritize these areas for renewable energy solutions. Telecom towers can provide anchor loads immediately and help improve economics of power distribution. Table 11.2 illustrates the various challenges of rural electrification and its solution by telecom infrastructure based model.

In case of any limitation or failure of RESCOs, similar subsidies can be extended to Telecom Infrastructure Companies or telecom service providers that can setup renewable energy base stations and supply excess power for community energy services. The business model for green power solution is illustrated in Figure 11.4.

Table 11.2: Power Solution for Telecom Towers in Rural Areas

Key Challenges for Rural Electrification	Solution by Telecom Infrastructure based Community Power Model
Power supply companies are not interested due to poor return	Mobile towers are situated even in remote areas. The RESCO can supply to mobile tower and surrounding areas.
Poor financial sustainability due to unclear revenue stream	Steady revenue stream from mobile operators can ensure financial sustainability
Poor operation and maintenance (O and M) of power equipments	High reliability requirements of base stations ensure proper O and M by telecom operator/RESCO
High transmission and distribution (T and D) losses	Mobile towers are located near to communities reducing T and D losses

Source: GSMA, Green Power for Mobile..

Advantages of Proposed Business Model

It will be a win –win situation for all *i.e.* telecom infrastructure provider and rural community. The advantages can be summarized in the following ways:

Ensure uninterrupted power supply to the telecom equipment:

  ✵ Provide clean, green, and reliable, pollution free, low emission and distributed technology power to telecom tower.

  ✵ Saves from high-running cost of generator and increasing diesel cost

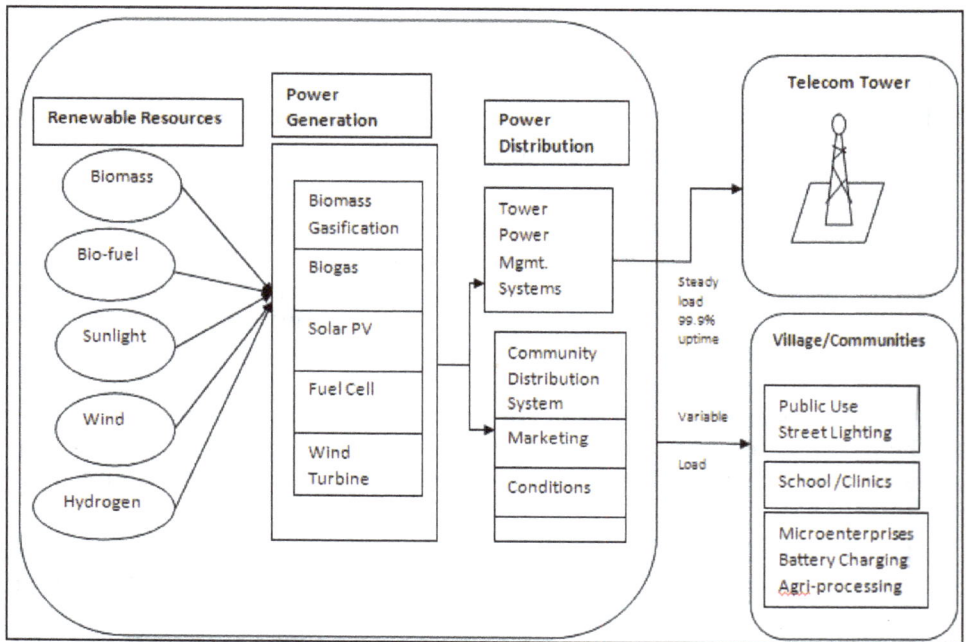

Figure 11.4: Business Model for Green Power for Rural Telecom.

★ The renewable energy system gives quality power output of 48 volt DC to charge directly the storage battery or provide direct power to telecom installations.

★ Efficient and easy installation, longer life.

★ Low gestation period and low operating cost

Conclusions

The proposed model will help in climate change mitigation, electrification of non grid connected area. It will be a business case for telecom operators to be part of this green telecom movement as it will help in increased revenue by getting new subscribers in rural area.

References

1. Gerhard Fettweis Ernesto Zimmermann (2008). ICT Energy Consumption – Trends and Challenges. In The 11th International Symposium on Wireless Personal Multimedia Communications (WPMC 2008). Retrieved from https:// mns.ifn.et.tu dresden.de/Lists/nPublications/Attachments/559/Fettweis_G_ WPMC_08.pdf

2. GSMA (2014). Green Power for Mobile. Retrieved from http://www.gsma. com/mobilefordevelopment/wp-content/uploads/2015/01/140617-GSMA-report-draft-vF-KR-v7.pdf

3. Kumar Amit, Lu Yunfei, Sood Manu and Singh S G (2010). Sustainability in Wireless Mobile Communication Networks through Alternative Energy Resources. *IJCST* Vol. 1, Issue 2, December 2010.

4. Nema Pragya, Nema R K, Ragnekar Saroj (2010). PV-solar/wind hybrid energy system for GSM/CDMA type mobile telephony base station. *International Journal of Energy and Environment*, Volume 1, Issue 2, 2010 pp.359-366.

5. Lange, C. Kosiankowski, D., von Hugo, D. (2015). Analysis of the energy consumption in telecom operator networks. Photonic Network Communications. doi:10.1007/s11107-015-0492-4

6. Matthews, E. P., Nici, J., Polonsky, B., and Wisniewski, J.F. (2010). Operating in the green: Modeling eco-friendly telecom network management. *Bell Labs Technical Journal*, 15(2). 175–192. doi:10.1002/bltj.20448

7. Sutherland, E.(2009). Climate Change: the Contribution of Telecommunications. *Communications and Strategies*, 4th Q.(76), 61–76.

8. Sutherland, E. (2011). Telecommunications and climate change: African and European experiences and requirements. SSRN, (July), 1–17. Retrieved from www.ssrn.com/abstract=1894605

Chapter 12

Sustainable Biohydrogen: A Candidate to Replace Carbon Based Energy

Khosrow Rostami*, Zahra Esfahani Boland Balaie
and Hassan Ozgoli

Iranian Research Organization for Science and Technology,
Tehran-33531-36846, Iran
*E-mail: rostami2002@yahoo.com

Abstract

Throughout, human history has been going with struggle for better standard of life. Energy is one of the key chain affecting source to begin with its, quality, safety to transport and storage being available with no international constrains, and environmentally friendly *etc.* Internationally published document is not yet available to recommend a reliable source to replace limited and polluting petroleum. However, energy source may have such potential provided is produced by sustainable resources such as microorganisms, algae, pyrolysis of biomass, electrolysis of water, wind, hydrolytic and solar energy *etc.* Sustainable biohydrogen are produced by dark and light fermentation. The data worldwide available regarding energy sources and types in published literature exhibits the need for modeling energy production in terms of supply and demands for international markets there are plenty in numbers and in varieties. It appears there is certain unsolved problem regarding economical hydrogen storage and transportation; however it would be resolved as the rest of technology is developed.

Keywords: *Biohydrogen production, Carbon energy, Greenhouse Gases (GHG), Hydrogen energy.*

Introduction

Dawn of the man history starts with burning of woody materials to generate light, energy, scaring of wild animals and warmth, as a result carbon monoxide and carbon dioxide and a lot more greenhouse gases have been added to the environment in addition to natural events. However, the gradual commercialization of man made products was milestone to pollute air, water and soil by industrialized nations. The major energy supply has come from exhaustible fossil fuels, which affect the climate change, have become a serious concern since late 1980s. One of the main issue is the increasing average world temperature because of the Greenhouse Gases (GHG) increment, and in particular the increased contribution of carbon dioxide (CO_2) emissions that contribute to the problem of global warming. Among the six kinds of GHG, the largest contribution to the greenhouse effect is of carbon dioxide (CO_2), and its share of greenhouse effect is about 56 per cent. The reduction of emitted GHG and atmospheric pollutants due to their adverse effect constitutes a foremost objective of contemporary energy and environmental policy in the world. In particular, the finding of the scientific community with respect to the rising of energy-related CO_2 emission not only shows the international awareness but also emphasizing the concerns about the issue. China is followed by the United States, India, Russia, Japan, Germany, Iran, Canada, South Korea, and United Kingdom that are the most emitters of GHG in the world. These nations are also the signatories to the United Nations Framework Convention on Climate Change (UNFCCC), and all the governments have approved the Kyoto Protocol in August 2002 and many others such as that of Copenhagen (2015) and very recently the Paris (2016). However, so far there is no substantial achievement, where the exception is Japan (CO_2 makes up 94 per cent of Japan's GHG emissions) and a few more nations have reduced GHG conservatively. Finally, it appears a strong alternative is required to replace the carbon energy source by a safer one. However, the question is not fully answered that is whether the wind, solar, geothermal and wave energy have all the criteria that petroleum fuels have. The answer may not be adequately encouraging; however hydrogen is a carrier of energy and could be produced through sustainable means such as production by bacteria in dark and light fermentation and biomass pyrolysis. Hydrogen contains energy to the level of 121.3 kJ g^{-1}, which is the highest among the present energy sources known. Hydrogen on combustion produces only water vapor that has half life of less than three hour, which is on the contrary beneficial to the life cycle. Further, when used in fuel cell environment has no adverse effect, as a result hydrogen economy may be one of the strong candidates to replace carbon economy. Nevertheless, it has to go a long way to be more economical, in terms of production, purification, compression, storing and transportation.

The data worldwide available regarding energy sources and types show the need for modeling energy production in terms of supply and demands for the universal scenario that is a difficult task and there are plenty in the published literature. In particular, there have been debates as to whether the methods based on the Divisia index are preferred to those based on the Laspeyres index, and vice versa. These are by far the two most popular decomposition approaches and in each case a number of different methods have been proposed by researchers. Interrelationships among energy, environment and economic welfare, and, simulated effects of several policies by using a 10-sector input–output model has been performed in the UK. Further, different methods have been adopted by international organizations, national agencies, researchers and analysts, and more often no method selection has been made on an adhoc basis indeed, there is no simple answer to the above mentioned questions. The Pundit of international era of energy should let the world to progress with a safer source of energy. In the United States, fossil fuels accounted for 86 per cent of total energy consumption in the year 2004. Petroleum fuels, natural gas, and coal accounted for 40, 23, and 23 per cent respectively, with additional 8 per cent from nuclear power and only 6 per cent from renewable sources, including hydroelectric (2.7 per cent), biomass/biofuels (2.7 per cent), and 0.6 per cent from solar, wind, and geothermal energy sources combined [11,12]. Currently available fossil fuel sources are estimated to become nearly depleted within the next century, with petroleum fuel reserves depleted within 40 years[11,13]. The United States imports 10 million barrels of oil per day of the existing world reserves (1.3 trillion barrels). Peak oil, the maximum rate of oil production, is expected to occur between 2010 and 2020.[11] Even with increasing attention on hydrogen as an alternative fuel, 95 per cent of worldwide production of hydrogen gas is from fossil fuel sources, primarily the thermocatalytic reformation of natural gas[14].

Efficiency of Fossil Fuels Used as Energy Source

The main fossil fuels (coal, natural gas, and oil) are about 33 per cent efficient when used for energy generation, and emit high levels of CO_2, and nitrogen oxides. Geothermal and solar energy are less than 20 per cent efficient with current technology, but are nearly zero-emission energy sources. Wind power has both high efficiency and zero-emissions, but is restricted to certain regions. Home heating by natural gas has a high efficiency, with lower emissions than other fossil fuels.

The reality of an eventual transition to hydrogen becomes more evident when one takes a realistic view of energy history. The world has been slowly shifting from one form of energy to another from solids to liquids to gases, since the mid-nineteenth century. Up to mid nineteenth century, focus on wood as energy was common in most settled parts of the world. However, in Great Britain, where population density and energy use were growing rapidly, wood began to lose out to coal, an energy source that was abundant such as wood but more concentrated, which was bulky and awkward to transport. Coal remained *Raja* of the world energy for the remaining of the nineteenth century and well into the twentieth. However, by the year 1900 the advantages of an energy system based on fluids, rather than solids, emerged as the transportation system started to shift away from railroads and toward automobiles. This shift created problems for coal, with its weight and volume, and at the same time it generated opportunities for oil, which featured a higher energy density and an ability to flow through pipelines and into tanks. By mid-century, oil had become the world's leading energy source[1]. But dominant as oil is, the liquid now faces an up-and coming challenger as gas. Despite improvements from wellhead to gasoline pump, the distribution of oil is rather cumbersome. Natural gas, in addition to being cleaner and lighter and burning more efficiently, can be distributed through a network of pipes, more efficient, and more extensive than the one used for oil. Natural gas has been the fastest-growing fossil fuel, the fuel of choice for electricity, and the second leading energy source, overtaking coal in 1999[1].

Hydrogen Production

Hydrogen exists in nature only in combination with other elements (with oxygen in water; carbon and oxygen in organic materials and fossil fuels) and is produced from its compounds, using energy sources. The benefits deriving from the hydrogen use depend, to a large extent, from the specific features of the production processes. In short, the use of hydrogen supports the development of a new and more sustainable energy system, with higher diversification of primary sources and reduction of greenhouse gas emission, on condition that in the long term it can be produced through sustainable processes. Hydrogen can be obtained from a wide variety of sources, such as fossil fuels, biomass, water, *etc.*, with different production processes.

Hydrogen Production from Fossil Fuels

Hydrogen production technologies from fossil fuels (steam reforming, partial oxidation, gasification) are mature and widely used (presently they provide more than 95 per cent of the hydrogen produced), even if they need to be optimized, for large-scale production, from the point of view of energy efficiency, environmental impact and, above all, costs. Furthermore, depending on the process and primary source used, the production from fossil sources, to be sustainable in the medium-long term, should be coupled with capture and storage of the co-produced CO_2. The development of solutions for cost-effective and reliable CCS is an essential condition in the long-term not only to produce hydrogen from fossil sources but, in general, to make possible the use of fossil fuels in a way compatible with the environment.

Hydrogen Production from Hydrocarbons

At present about 78 per cent of hydrogen is produced from fossil fuel, natural gas (48 per cent) and other hydrocarbons (30 per cent). The most economically advantageous process is the catalytic steam reforming of natural gas (but also liquefied petroleum gas or naphtha can be used in the process). Possible improvements of the process are mainly focused on recovery of thermal energy, process integration and gas purification.

The typical capacities of industrial plants range between 100,000 and 250,000 Nm^3h^{-1}, with thermal efficiencies of about 80 per cent. Medium (200-500 Nm^3h^{-1}) and small (50-300 Nm^3h^{-1}) capacity plants are suitable for specific industrial applications, with reduced efficiencies and higher costs per product unit. Systems of lower capacity (1-50 Nm^3h^{-1}) have been developed for integration with fuel cells and are currently used in hydrogen refueling stations; in this case, further improvements are needed, mainly related to the integration of different equipment, gas purification and study of innovative catalysts[2]. For heavier hydrocarbons, auto-thermal reforming or partial oxidation processes with thermal efficiencies higher than 75 per cent are used. The cost for hydrogen produced through steam reforming strongly depends on the plant size and the cost of natural gas (for large plants it represents 50-70 per cent of the final cost). For small sized plants the hydrogen cost is presently around 20 €/GJ; in 2020, taking into account the possible technology optimization and the increase of natural gas price, a slightly higher cost is expected[2-4].

Coal Gasification

At present, about 18 per cent of the hydrogen produced worldwide is derived from coal gasification in large-scale central facilities (100,000 - 200,000 Nm^3/h). Three types of gasifiers are available: moving (or fixed) bed, fluidized bed and entrained flow. The overall efficiency is about 60-65 per cent, with a reduction of 3-6 per cent points in plants where CO_2 capture and storage is provided. The technology is mature, even if it is more complex and less consolidated than steam reforming. The potential for further improvements is still considerable and takes into account innovative membranes for air separation, progress in gasifier configuration, hot gas purification systems, new solvents and membrane reactors for hydrogen separation, *etc.* Coal gasification, integrated into combined cycles with CCS, represents also a very interesting option for centralized cogeneration of electric energy and hydrogen. Such plants will have specific costs equivalent to those of similar plants designed only for electric energy generation and hydrogen costs equal or less than those of hydrogen production plants [5]. The cost difference between hydrogen produced from natural gas and coal could be lower in future, both because coal gasification has larger margins of improvement in the medium term, compared with steam reforming, and coal cost could be subject to cost rises lower than the ones of natural gas.

Production from Water Electrolysis

Electrolysis allows the splitting of water in its constituents, hydrogen and oxygen, using electric energy. The process has, at the present state, significantly higher costs than hydrogen production from fossil fuels and covers only a small share of the world production (4 per cent). The process is mostly used to satisfy requests for high purity hydrogen. The electrolysers in the market are essentially of two types: alkaline electrolysers, that use an aqueous solution of potassium hydroxide (KOH), and solid polymer electrolysers, where the electrolyte is a polymer membrane (the same as the polymer electrolyte fuel cells)[6]. The solid polymer electrolysers present some advantages (absence of corrosive liquids, higher current density and operating pressures), but have durability problems (particularly for the limited lifetime of some membranes). The factor that mostly influences the hydrogen cost is the system efficiency, considering that the electric energy cost represents, for plant of significant sizes, about 80 per cent of the hydrogen cost. It is estimated that, for large plants, the hydrogen cost from electrolysis could be reduced from 30 €/GJ in 2010 to 15 €/GJ in 2030 (with an electricity price of 0.03 € $kW^{-1}h^{-1}$)[7]. The cost of renewable hydrogen production is presently a lot higher than the one from electric grid, even if a significant reduction can be expected in the medium term. These technical objectives (added to the energy and economical ones) have remarkably widened the set of solutions and materials studied with the support of public funding; in the U.S. Department of Energy Program about 30-35 M€ have been provided, while the European Commission invested in total about 25 M€ in the Sixth Framework Program.

Hydrogen Production from Renewable Sources (Biomass and Bacteria's)

Hydrogen can be produced from biomass using different thermochemical (gasification, pyrolysis) and biological processes. Among the thermochemical processes in general, gasification is more suitable for centralized production. The process is similar to that used to produce hydrogen from coal. The biomass can have different origin, such as agricultural and forest residues, industrial and urban wastes, organic waste materials, *etc.* The anaerobic digestion of organic materials is at the moment the most promising biological process for hydrogen production. This process, tested at laboratory scale, has a theoretical specific reactor yield of 10-20 m^3day^{-1} of hydrogen and a substrate consumption of about 20-100 $kgm^{-3}day^{-1}$. The technology offers considerable potentials, but a large research effort is still needed to understand and optimize the whole process in order to proceed to its scale-up [8].

Photo-Biological Production

Hydrogen can be produced from water using sunlight and photosynthetic microorganisms. The process has reached at laboratory scale interesting conversion efficiencies (about 2 per cent of the incident light radiation), even if it still requires important improvements, both to understand the basic mechanisms of process and for its scale-up. Photo-biological water splitting is a long-term technology.

Biological Production of Hydrogen

Biological production of hydrogen can be carried out by two main mechanisms: photobiological process and fermentation [9,10]. Photobiological hydrogen production has the advantage that it utilizes solar radiation; to drive the process advanced bioreactor designs are required to achieve moderate solar radiation conversion efficiency and H_2 production rates.

Fermentation processes can utilize free carbon/energy sources in agricultural by-products or wastes, but not all fermentative microbes are contained in many feed stocks. Therefore, both photobiological and fermentation hold great promises as future contributors to global H_2 production, but technical challenges remain that need to be overcome.

Biological hydrogen production carried-out in photobiological or fermentations paths are dependent on hydrogenase or nitrogenase enzymes that catalyze the reduction of protons (eq. 1.).

$$2H + 2e^- \rightarrow H_2 \tag{1}$$

In general, hydrogenase enzymes can catalyze the reaction in either direction in vitro. In vivo, they primarily catalyze either production or uptake (oxidation) of hydrogen depending on the needs of the host microorganisms. In fermentative bacteria, hydrogenase enzymes catalyze the reduction of protons to molecular hydrogen as a means of accepting extra electron produced from the oxidation of organic substrates, while for photosynthetic bacteria, hydrogenases catalyze the use of hydrogen as an electron donor[11]. Until today several classes of enzymes have been identified: nitrogenases, Fe-hydrogenases, NiFe-hydrogenases (including NiFeSe hydrogenases), and nonmetal hydrogenases.

Photobiological Hydrogen Production

Photobiological hydrogen production has great potential because the energy available in solar radiation is very large. Yearly average solar irradiation can be as high as 5 kWhm^{-2}day^{-1} or 6.6 GJyear^{-1} (10^9 J year^{-1})[18]. At a high solar radiation conversion efficiency of 10 per cent and at a price of $15GJ^{-1} for H$_2$, light driven H$_2$ production can yield $10/m^2-year [18]. The main limiting factor for all light driven processes are the low conversion efficiency of high intensity solar energy. Under full sunlight conditions, in oxygenic (oxygen generating) photosynthesis, the rate of light energy capture by photosynthetic pigments is 10 times greater than the rate of transfer of electrons from Photosystem II to I. This means that ~90 per cent of solar radiation captured is not utilized in photosynthesis; rather, it is released primarily as heat. To avoid light saturation, advanced reactor designs may use rapid mixing to take benefit of the "flashing light effect" that is the rapid exposure of algal cells to maximum light capture. Therefore the goal of much H$_2$ production research is to increase solar energy conversion efficiencies beyond the 3 per cent currently achievable with outdoor microalgae cultures at high solar intensity. The photo-dependent hydrogen processes include direct biophotolysis, indirect biophotolysis, and photofermentation [17,18].

Direct Biophotolysis

In oxygenic photosynthesis, two photosystems are used. Light energy that strikes Photosystem II (PSII) is used to split water molecules generating oxygen, protons, and electrons (eq. 2.):

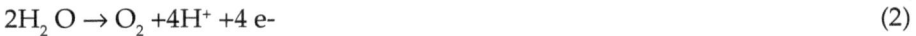

$$2H_2O \rightarrow O_2 + 4H^+ + 4 e- \tag{2}$$

The electrons are accepted by PSII, reducing it to a potential of approximately -0.8 V. The electron then flows through a series of carrier molecules to Photosystem I (PSI). Acceptance of the electrons at PSI reduces it to a potential that is sufficient to reduce ferrodoxin, which in turn reduces NADP to NADPH. In the typical process for photoautotrophic growth, the NADPH is then used to reduce inorganic carbon for synthesis of new cell mass. This reaction is ubiquitous in algae and cyanobacteria, so its potential for energy production is great if it could be fully exploited. In direct biophotolysis, an alternative route for the electrons is utilized. The electrons transferred to ferrodoxin in PSI are transferred to protons instead of NADPH, reducing the protons to hydrogen gas through the activity of Fe-hydrogenase enzyme. The Fe-hydrogenase enzymes needed to catalyze this reaction are found in the chloroplasts of green algae such as *Scenedesmus obliquus*, *Chlamydomonas reinhardtii*, and *Chlorella fusca* and in many *cyanobacteria*. The key problem with direct biophotolysis is that the Fe-hydrogenases that catalyze the reduction of H$^+$ are extremely sensitive to inhibition by oxygen. Thus, direct biophotolysis requires photobioreactor designs that allow for production, capture, and separation of H$_2$ and O$_2$.

Despite innovative photobioreactors designed to maximize H_2 production and attempts at molecular engineering of hydrogenases that are tolerant of oxygen, obstacles to efficient hydrogen production by direct biophotolysis remain[21].

Indirect Biophotolysis

The process of indirect biophotolysis is accomplished by separating H_2 and O_2 productions both temporally and spatially. In the first step, light energy is used to produce oxygen and stored as carbohydrate. Typically starch in green algae and glycogen in cyanobacteria imposing nitrogen limitation enhances accumulation of large amounts of stored carbohydrate in both the green algae and cyanobacteria and results in loss of oxygen evolving capacity. In the second step stored carbohydrate is converted to H_2 and CO_2 in light driven processes and under oxygen-depleted conditions:

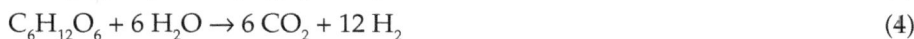

$$6 CO_2 + 6 H_2O \rightarrow C_6H_{12}O_6 + 6O_2 \tag{3}$$

$$C_6H_{12}O_6 + 6 H_2O \rightarrow 6 CO_2 + 12 H_2 \tag{4}$$

Cyanobacteria, such as Aphanocapsa montana, Anabaena variablies,and Spirulina produce hydrogen through indirect processes[17,18]. Alternatively, the stored carbohydrate could be fermented to 4 mole of hydrogen gas per mole of stored glucose in a dark anaerobic fermentation process. One proposed strategy would utilize a four-step process[10,18]: (1): Photosynthetic production of algal with high stored carbohydrate content and a potential of 10 per cent solar energy conversion in a low cost open raceway systems (2) Concentration of algal biomass (3) anaerobic, dark fermentation of stored carbohydrate in concentrated algal biomass to hydrogen gas and acetate (4 mole H_2 and 2 mole acetate are produced per mole of fermented glucose); and (4): final phase utilizing closed photobioreactors for indirect biophotolysis to use light-driven conversion of remaining stored carbohydrate and acetate produced by fermentation to hydrogen gas.

Photofermentation

Photosynthetic bacteria such as *Rhodopseudomonas capsulates*, or *R. sphaeroides*, and other non-sulfur bacteria have the capacity to produce H_2 through the action of nitrogenase enzyme in a process called photofermentation. In this processes organic materials are oxidized under anaerobic condition using light energy when nitrogen is limited. Excesses e- is transferred to ferroxidin and then to protons, producing H_2 gas as catalyzed by nitrogenase. The H_2 producing activity of nitrogenase is simulated by light, but is inhibited by ammonium and nitrogen gas,and requires substantial amount of energy. Hydrogenase enzymes, which play a critical role in H_2 production in direct biophotolysis, primarily act to consume H_2 in photofermentation. Mutated cells with reduced production of hydrogenase results in higher hydrogen production. Organic substrates such as lactate or malate result in greater H_2 production rates than sugars such as glucose or sucrose employing batch conditions. Therefore, conditions required for optimum H_2 production by photofermentation include maximal nitrogenase activity, minimal hydrogenase activity, use of favorable substrates such as lactate or malate, high carbon to nitrogen ratio in media and light distribution throughout the culture depth.

Photobiological Hydrogen Production Potential

Because of the low solar energy conversion of all photobiological H_2 production, Hallenbeck and Benneman[5] conclude that continued basic research and development are needed before photobiological hydrogen production can be fully utilized, and that conversion of organic substrates to H_2 through dark fermentation processes is more efficient and has greater potential for expanding biohydrogen production.

Hydrogen Production by Fermentation

Hydrogen gas is produced under anaerobic conditions by certain chemoorganotrophic microorganisms that use organic substrates as their carbon and energy source and hydrogen ion as electron acceptor. The biological production of hydrogen by fermentation is generally associated with the presence of an iron-sulfur protein called ferredoxin, an electron carrier of low redox potential. Hydrogen can be produced by mesophilic (25 to 40°C), thermophilic (40 to 65°C), extreme thermophilic (65 to 80°C), and hyperthermophilic (>80°C) microorganisms of the *Archaea* and Bacteria Domains. In the hyperthermophilic group, most of the H_2-producers are *Archaea*, but one group, the *Thermotogales*, are bacteria. In addition to H_2, cofermentations result in the formation of a variety of organic acids such as acetate, butyrate, lactate, or propionate, along with lesser amounts of alcohols such as ethanol. The amounts and rates of formation of each of these products depend on the organism and substrate concentration, as well as culture condition, pH, and temperature.

Hydrogen Storage

The problems related to hydrogen storage derive from its chemical-physical characteristics. Hydrogen is a fuel that has a high gravimetric energy density, but also low volumetric energy density, both at gaseous and liquid states. Consequently it is clear that, compared to other fuels, hydrogen requires higher volume tanks to store the same energy content. The storage system should have high density of energy (corresponding to high amounts of stored hydrogen), high density of power, good energy efficiency, low boil-off losses in liquid hydrogen storage, adequate life-cycle, reduced or no environmental impact and acceptable safety features (both during operation and in the phases of manufacture and disposal at end of life), reduced costs and efficient, fast and safe filling at the refueling station. Different technologies are already in use or under development for hydrogen storage. It can be stored as high pressure gas, in liquid or chemical form, absorbed/adsorbed on special materials (metallic hydrides, chemical hydrides, carbon nanostructures). Each technology shows advantages and limitations, but all of them, even where they are already applied, require still significant R&D efforts for a reliable and competitive large scale use. The current ambitious research targets aim to develop and demonstrate hydrogen storage systems achieving 2 kWhkg^{-1} (6 wt per cent H_2) and 3 kWhkg^{-1} (9wt per cent H2) [9–10].

Hydrogen Utilization

Hydrogen, in addition to the traditional industrial applications, can be used in both transport (internal combustion engines and, above all, fuel cells) and stationary power generation (thermal cycles and fuel cells) sectors.

Hydrogen Vehicles

Hydrogen use in transport allows to obtain zero (with fuel cells) or very reduced (with internal combustion engines) emissions at local level and, depending on the source used for hydrogen production, can give a significant contribution to the reduction of vehicle greenhouse gas emissions. Vehicles using Internal Combustion Engines (ICE) can operate with both pure hydrogen or hydrogen-natural gas blends. Significant experiences on the use of hydrogen have been obtained with modified conventional engines. To exploit the potential advantages of hydrogen at the best it is however necessary that engines are designed taking into account its characteristics as fuel (wide flammability range in comparison with other fuels, low ignition energy and almost double flame rate). The main car manufacturers are confident that the realization of hydrogen engines is feasible using the currently available technologies, once the boundary conditions (creation of fueling infrastructures, development of regulations and standards) allow to generate a sufficient market volume. Testing ICE vehicles (cars or buses) fed with hydrogen has been carried out or is in progress in USA and Europe, even if the commitment is significantly lower than that for fuel cell vehicles. Car manufactures such as BMW, Ford Motors and Mazda are involved in the development of hydrogen-fed ICE technology. Another solution under evaluation contemplates the use of blends of natural gas and hydrogen (HCNG), at variable content of hydrogen, but in any case not higher than 30 per cent by volume, to avoid engine modifications. The addition of hydrogen to natural gas, although at low percentages, has positive effects on engine operation, reducing exhaust emissions, not only due to the substitution of the carbon with hydrogen, but also because its presence allows a more complete and rapid combustion, with a significant efficiency increase [11].

Fuel Cell Vehicles

The use of fuel cell systems powered with hydrogen represents one of the most promising options in the medium-long term for the development of efficient and environment-friendly means of transport, as the use of hydrogen implies zero emissions at local level. Besides fuel cell vehicles offer efficiencies almost two times higher than conventional vehicles, maintaining similar performances in terms of driving range, top speed and acceleration.

Among the various fuel cell types, the Polymer Electrolyte Fuel Cells (PEFC) are the most suitable technology for transport applications, being characterized by low operation temperature, high power density, quick start-up and rapid response to load changes. In the last decade, the governments of the most industrialized countries have started R&D programs in this field, investing several million dollars, and major auto manufacturers are actively engaged in fuel cell vehicle development and demonstration, with high resource commitments (it is estimated that Daimler and General Motors have invested more than 1 billion dollars in the last years). Some car industries (General Motors, Honda, Nissan and Toyota) have in-house programs to produce their own fuel cell stacks, while others (Daimler AG, Ford Motor) prefer to make supply agreements with fuel cell developers such as Ballard Power Systems, UTC Power or Nuvera Fuel Cells. Despite substantial progress, short-medium term R&D efforts are still necessary before fuel cell vehicles become widely available and achieve significant market share. It is needed, in fact, to overcome a number of technical and economic barriers. In addition to the availability of a wide hydrogen transport and distribution network and reliable and safe technologies to store it on-board, it is necessary to reach a stack durability of about 5,000 hours and a good reliability, but above all to achieve costs compatible with the transport market. Being the cost of an ICE of 20-30 €/kW, a fuel cell system to be competitive should cost less than 30-50 €/kW. Today, with low volumes of production and components built through manual fabrication processes, costs are around 2000-3000 €/kW. These costs are highly dependent on the number of vehicles on the road.

Possible Hydrogen Scenarios

In the last years many studies [12,5,8] have tried to outline possible scenarios for hydrogen development as an energy carrier. A synthesis of the results of these studies was provided by the European Roads2HyCom project [12]. All the analyses show that hydrogen will be able to have a significant role in the energy system on the condition of strong environmental policies, high fossil fuel costs and adequate improvement of the involved technologies. For hydrogen applications, the scenarios do not take into consideration its use in the electric power generation, which could have an important role, in connection with the use of IGCC plants with CO_2 sequestration. However this possibility will only increase the importance of hydrogen in the energy market, strengthening its role.

In the desired scenarios, the road transport sector will use about 75 per cent of hydrogen. The main technology will be fuel cell vehicles, even if ICE vehicles will have a significant market share, especially at the beginning. The first batch vehicles will be introduced in the public transport fleets, both for technical reasons (reduced problems for on-board storage and distribution infrastructures), and because their wide-scale use could be supported from public contributions, on the basis of their environmental benefits. Niche markets (small electric vehicles, forklifts, boats) will have an important role in the first phase as they will be able to contribute to the cost reduction and the creation of the distribution network. A significant introduction in the private car market is not envisaged before 2020 (at such time the European Platform assumes a market share between 1 and 3 per cent).

Four possible HyWays scenarios, based on different assumptions about the evolution of technology and the supporting policies are carried out in the HyWays project [8]. In 2050 the most optimistic scenario shows that hydrogen vehicles represent about 75 per cent of the passenger cars, while in more conservative scenarios the hydrogen share would not be higher than 30 per cent.

Greater international collaboration in supporting hydrogen is also needed. Twelve industrial nations are cooperating on hydrogen efforts under the auspices of the International Energy Agency (IEA). Under the agency's Hydrogen Implementing Agreement, created in 1977 to increase hydrogen's acceptance and wide use, the IEA has funded numerous research and development efforts and demonstration projects. The program is geared towards a hydrogen future with sustainable energy, and thus focuses on solar production, metal hydrides, and the integration of renewable energy and hydrogen systems. It is also working to engage other interested countries, like China, Iceland, and Israel[22]. Hydrogen has stronger political support in Germany, which is the world leader in terms of the number of demonstrations of hydrogen and fuel cell vehicles, fueling stations and renewables-based hydrogen production systems, as well as in the hosting of hydrogen conferences. The German government recognizes that hydrogen is critical to its long-term energy strategy, and is expected to make the fuel a higher priority in coming months. However, hydrogen expert Dr. Rolf Ewald contends that federal and EC funding for hydrogen is "decreasing and weak," with the most support coming from German states such as Bavaria[23].

Japan's national program is considered the most ambitious and comprehensive of the world's hydrogen initiatives to date. Japan expects to spend about $4 billion on its WE-NET (World Energy Network) program by 2020. Currently funded at $88 million over five years, the program is involved in improving the efficiency of fuel cells, enhancing the storage capacity of metal hydrides; installing filling stations that will test out natural gas reformers and electrolysis; and testing cars using metal hydrides and compressed gas cylinders in partnership with Japanese automakers. Its scientists view natural gas reforming and electrolysis as the near-term infrastructure path, and hydrogen from renewable energy as the medium- to long-term route. However, WE-NET official Kazukiyo Okuno acknowledges that the program has not set any goals for introducing hydrogen into the market[24].

Currently the cost of hydrogen is more than twice as much as that of diesel and petrol on per unit energy basis and substantial progress is needed to make environmentally sustainable hydrogen production pathways cost-competitive with petroleum fuels, even assuming that hydrogen powered vehicles will be significantly more efficient than their conventional gasoline-powered counterparts [25-26].

Conclusions

On the other side, although technical and economic aspects are important for hydrogen deployment, it has to be underlined that they are not sufficient for the success of its technologies. In fact, as hydrogen technologies have been introduced mainly to face environmental and energy security issues, just lowering hydrogen costs could not be enough. The reason is that the environmental and energy security aspects can hardly be handled through merely economical penalties, although this approach has been chosen in the past, imposing a cost for CO_2. The only application of penalties could be misleading as they should be calculated taking into account the costs to be incurred to recover from the negative effects provided by a bad use of energy resources. However, if the degradation is not stopped soon, certain critical planetary conditions could be largely violated and there would be no chance to really recover from climate change related effects, even allocating very large resources. Therefore hydrogen deployment should be worthwhile only if other measures and criteria are promoted (*i.e.* better use of energy resources, increase of energy efficiency, decreasing non essential activities, *etc.*). Even in this case the effect will be really successful only if the new approach is pursued in conjunction with important changes of present human behavior, especially considering the most developed countries. It is therefore required that human beings modify their present life style substantially, privileging collective advantages instead of personal ones, to allow a large penetration of hydrogen and more benign technologies. This change is of course very complex and will require long and highly coordinated actions to be undertaken. Policy makers should take the lead among all the stakeholders providing suitable measures to foster hydrogen deployment and creating the conditions to increase public awareness through education, formation and information. It is clear at the end that hydrogen can be successfully deployed in the future only if a radical change is achieved of the way energy is perceived in human life. This means that the importance of energy for growth of society is fully understood by most of the citizens. More attention on importance of energy could also lead to important changes in the way energy is presently produced, distributed and consumed. Under this aspect hydrogen could play a significant role not only in transport applications, but also in residential and commercial uses. In fact hydrogen, especially if converted in fuel cells, can provide end users with locally produced electric energy and heat, modifying the present situation where a centralized electricity production is provided.

References

1. Ausube,J.H. (1999). Figure 1 from Robert A. Hefner, GHK Company, The Age of Energy Gases, adapted from presentation at the 10th Repsol-Harvard Seminar on Energy Policy, Madrid, Spain, 3 June 1999 (Oklahoma City, OK: 1999); Jesse H. Ausubel, "Where is Energy Going?"The Industrial Physicist, February 2000, pp. 16–19.

2. Galli, S.; Calò, E.; Monteleone, G. (2008). *Idrogeno da idrocarburi*; RT ENEA.

3. Mueller, L.F.; Tzimas. E.; Kaltschmitt. M.; Peteves, S. (2002). Techno-economic assessment ofhydrogen production processes for the hydrogen economy for the short and medium term. *Int. J.Hydrogen Energ.*, 32, 3797-3810C.

4. Benneman, J.R. (2000). Hydrogen Production by Algae. *J. of Applied Phycol.* 12: 291–300.

5. Hallenbeck, P.C., J.R. Benemann (2002). Biological hydrogen production; fundamentals and limiting processes. *Int. J. of Hydrogen Energ.* 27: 1185–1193.

6. Vignais, P.M., B. Billoud, J. Meyer (2001). Classification and phylogeny of hydrogenases. *FEMS Microbiol. Rev.* 25(4):455–501.

7. Adams, M.W.W. (1990). "The metabolism of hydrogen by extremely thermophilic, sulfur-dependent bacteria." *FEMS Microbiol. Rev.* 75:219–237.

8. Energy Information Agency (2007). "Official Energy Statistics from the U.S. Government." Energy Information Administration. Available at: http://eia. doe.gov/.

9. U.S. DOE (2004). "U.S. Energy Consumption by Energy Sources, 2000–2004."Available at: http://www.eia.doe.gov/cneaf/solar.renewables/page/trends/Table 01.pdf.

10. BP (2005). "Putting Energy in the Spotlight. British Petroleum Statistical Review of World Energy." Available at: http://www.bp.com/liveassets/bp_internet/switzerland/corporate_Switzerland/STAGING/local_assets/downloads_pdfs/s/statistical_review_of_world_energy_2005.pdf.

11. Sperling, D., J. S. Cannon (eds.) (2004). The Hydrogen Energy Transition: Moving Toward the Post Petroleum Age in Transportation. p. 80

12. Watkiss, P.; Hill, N. (2002). The Feasibility, Costs and Markets for Hydrogen Production; AEA Technology, Final Report, Sep.

13. World Energy Technology Outlook 2050 (2006). WETO-H$_2$, EU 22038.

14. European Hydrogen and Fuel Cell Technology Platform. Available online:https://www.hfpeurope.org/.

15. European Parliament legislative resolution on the Proposal for a Directive of the European Parliament and of the Council on the promotion of the use of energy from renewable sources (COM) 2008, 0019, T6-0609/2008; Available online: http://www.europarl.europa.eu/.

16. HyWays, The European Hydrogen Roadmap (2007). Available online: http://www.hyways.de

17. An investigation into the H$_2$ 5 per cent 2020 target of the European Commission; NOVEM-TUWien, Cleaner Drive Project, June 2004.

18. Hydrogen, Fuel Cells and Infrastructure Technologies Program Multi-Year RD and D Plan, U.S.Department of Energy, Oct. 2007; Available online: http://www.eere.energy.gov/Hydrogen and fuel cells/mypp/.

19. Ortenzi, F.; Chiesa, M.; Scarcelli, R.; Pede, G. (2008). Experimental tests of blends of hydrogen and natural gas in light-duty vehicles. *Int. J. Hydrogen Energ.*, 33, 3225-3229.

20. Prospects for Hydrogen and Fuel Cells; Int. Energy Agency, OECD Publishing: Paris, France, December 2005. Available online: http://www.iea.org/textbase/nppdf/free/2005/hydrogen2005.pdf.

21. Snapshots of Hydrogen Uptake in the Future. A Comparison Study, Road2HyCom Project, September 2007; Available online: http://www.roads2hy.com.

22. Neil Rossmeissl, US DOE, "The International Energy Agency's Hydrogen Research and Development Activities," in Forum für Zukunftsenergien, ed., op. cit. note 52, pp. 453–59; David Haberman, "Implementing a Practical Vision of the Hydrogen Economy," Presentation to Micro-power 2001 Conference, San Francisco, CA, 20 February 2001.

23. Jurgen Hansen, German Hydrogen Association, "Hydrogen Efforts in Germany and Europe,"in NHA, op. cit. note 53, pp. 349–65; Dr. Rolf Ewald, Deutscher Wasserstoff-Verband,"Hydrogen in Germany," in Forum für Zukunftsenergien, ed., op. cit. note 52, pp. 153–60.

24. Akihiko Ishikawa, MITI, "Hydrogen in Japan," in Forum für Zukunftsenergien, ed., op. cit.note 52; Yoshitaka Tokushita, New Energy and Industrial Development Organization,"Plan/Overview of the WE-NET (World Energy Network) Project," in Forum fürZukunftsenergien, ed., op. cit. note 52, pp. 559–60; Kazukiyo Okano, Director of Research, WENET Center, "WE-NET Phase 2 Program Update," in NHA, op. cit. note 53, pp. 379–87.

25. DuPont V. Steam reforming of sunflower oil for hydrogen gas production/ OXIDACIÓNCATALÍTICA DEL ACEITE DE GIRASOL. EN LA PRODUCCIÓN DEL GASHIDRÓGENO/REFORMAGE ÀL AVAPEUR DEL'HUILE DE TOURNESOL DANS LA PRODUCTION DE GAZ HYDROGÈNE. Helia 2007;30:103–32.

26. Acar, C., Dincer, I. (2013). Comparative environmental impact evaluation of hydrogen production methods from renewable and nonrenewable sources. In: DincerI, Colpan CO,Kadioglu F, editors. Causes, impacts and solutions to global warming. Springer; p. 493–514.

Chapter 13

The Resilience of Critical Urban Infrastructure Systems and Energy Efficiency: An Indian Perspective

Geeta

Research Associate,
Centre for Science and Technology of the Non-Aligned and
Other Developing Countries, (NAM S&T Centre),
New Delhi, India
E-mail: geetageet01@gmail.com

Abstract

Extreme climate change and emerging urbanization are making cities more endangered to loss of electric power and damage to energy infrastructure. The efficiency and conservation of energy is a major concern worldwide, given modern society's strong dependence on its adequate delivery. Not only does the functioning of industry, transportation, and communication and computer systems depend on continuous energy supply, but our complete style of living collapses when energy fails. Surges in fuel prices, wars and natural disasters directly impact on energy supply. The occurrence of natural disasters and their impact on electric power system functioning is also of concern to the countries. Energy consumption in transport sector is also increasing along with an increase in road traffic and air travel. These trends will lead to increased transport energy use, leading to higher rate of CO_2 generation. To meet the ever growing worldwide energy consumption demand, efficient technologies are being developed globally and deployed in various sectors. In India, it is anticipated that the steady increase in GDP and improving standards of living will continue to drive energy consumption in these sectors. The transport sector in India is also the major contributor to GHG emissions. Indian cities are densely populated with poor infrastructure which is more vulnerable to climate change. Reduction of GHG emissions and increased energy efficiency through intelligent and inclusive infrastructure alone can make our cities climate-resilient.

In order to find out the resilient solutions for sustainable development and energy conservation, this paper lists down the existing policy framework and focuses on the potential of GHG reduction and energy efficiency opportunities enabled through ICT and resilient infrastructure solutions with particular reference to an Indian perspective.

Keywords: *Climate change, Energy efficiency, Information and Communication Technology (ICT), Greenhouse Gas (GHG), Resilient-sustainable infrastructure.*

Introduction

Rapid population growth, large scale environmental change and a globalized economy make today's world one of increasing complexity, uncertainty and continuous transformation. Directly and indirectly, these factors give rise to the growing frequency, magnitude and geographic range of major stresses. In cities, the challenges are especially acute. People, infrastructure and economic activity amass in urban areas, concentrating high value in often exposed locations. Today, cities generate more than 80 per cent of global Gross Domestic Product (GDP). Cities have always faced risks, but in a time of rapid change – including a changing climate – cities need to plan for and manage risks differently from before, in a way that includes strategies to deal with complexity and uncertainty. Considering the impact of climate change on the country, it is imperative to undertake nationally appropriate actions to contend with the consequences. This approach is at the heart of resilience planning and energy efficiency.

Resilience thinking is about generating increased knowledge on how we can strengthen the capacity to deal with the stresses caused by environmental change. Resilience works within the context of long-term sustainability objectives but specifically embraces the turbulence of daily life. Resilience is about learning to live with the spectrum of risks that exist at the interface between people, the economy and the environment, and maintaining an acceptable stability or equilibrium in spite of continuously changing circumstances. Resilience also addresses the interdependencies between systems and minimizes unforeseen 'gaps' in risk and energy management. At times, a resilience approach may appear contrary to accepted definitions of sustainability. For example, while sustainability may encourage leaner, more efficient operations in the interests of resource conservation, resilience promotes greater redundancy in city infrastructure to provide back-up during a crisis. Such tensions can be an important signal that short term efficiency gains may not in fact be the right pathway to long term sustainability. Therefore, system planning and design should seek to measure performance against both resilience and sustainability indicators. Many of the technologies presented in this paper demonstrate the possibility of achieving energy efficiency and sustainable development simultaneously. Through this study, innumerable opportunities have been identified to deploy ICT and Resilient-Sustainable Infrastructure solutions towards accomplishment of these mitigation and energy efficiency related objectives.

Energy Consumption and Emission in Transport and Electricity Sectors in India

(i) Transport

India's transport sector is large and diverse and this sector feeds to the transport needs of 1.3 billion people. The steady increase in per capita GDP continues to drive increasing demand for mobility and vehicle use in India. The transport sector is a major contributor to air pollution as well as CO_2 emissions in densely populated urban India. Passenger mobility in urban India relies heavily on its roads, as rail-based and air based transport services are available in only select cities. The transport sector in India consumes about 16.9 per cent (36.5 mtoe) of the total energy consumption (217 mtoe in 2005–2006). Various energy sources used in this sector are coal, diesel, petroleum (gasoline) and electricity. Transportation through road, rail and air are responsible for CO_2 emissions of around 80 per cent, 13 per cent and 6 per cent, respectively. Historically, the rapid increase in number of vehicles and travelling distance has resulted in higher consumption of energy, with an average annual growth rate of 2.9 per cent [1]. The urban population of India, which constitutes 28 per cent of the total population, is predominantly dependent on road transport. In India, around 80 per cent of passenger and 60 per cent of freight movement depends on road transport [1]

(ii) Electricity

In an indication of growing appetite for electricity in India, the country's per capita electricity consumption has reached 1010 kilowatt-hour (kWh) in 2014-15, compared with 957 kWh in 2013-14 and 914.41 kWh in 2012-13 [2]. The current power infrastructure in India is not capable of providing sufficient and reliable power supply. Another problem is unstable power supply. There are frequency fluctuations caused by load-generation imbalances in the system and this keeps happening because consumer load keeps changing. Frequency is the most crucial parameter in the operation of AC systems. The rated frequency in India is 50.0 Hz. While the frequency should ideally be close to the rated frequency all the time, it has been a serious problem in India. Poor power quality control has knock-on effects on equipment operation, including large-scale generation capacity. Challenges do exist in the sector, which India has to overcome to evolve from a developing market to a matured market. Meanwhile, the gap between what can be achieved and what is currently present, uncovers a number of possibilities and opportunities for growth. Large percentage of power produced in the country is lost to inefficiencies in the state distribution networks. Indeed, distribution is the weakest link in India's power story, with a loss figure that stood at 38.86 per cent in 2000-01. More than 30 per cent of the power produced in the country is lost to theft and inefficiencies. This is nearly the same load that the northern, eastern and the north-eastern grids were carrying when they tripped in Aug. 2012 putting millions in darkness. By applying the structured energy description of a grid as a topology of micro grids, including nesting, union and intersection, we demonstrate a new way to understand fault resilience, containment, and recovery. Factoring operations management into micro grid management and cross-micro grid management reduces the complexity of each.

Figure 13.1 shows the different sources of electricity generation in India which indicates that about 57 per cent of the electricity is generated from coal, 19 per cent by hydroelectric power plants, 12 per cent by biomass and other renewable, 9 per cent from gas, and 2 per cent from nuclear power plants. Thus India's commercial energy demand is met through the country's vast coal reserves.

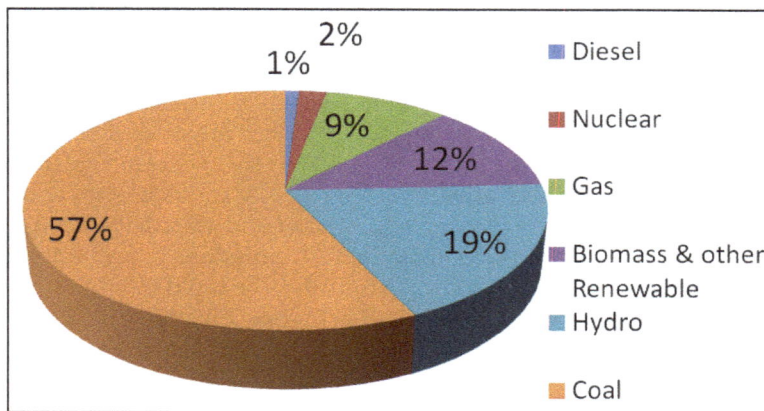

Figure 13.1: Electricity Generation Sources in India (*Source*: EIA [19]).

Resilient Solutions for Enhancing Energy Efficiency

Transport Sector

There are different kinds of measures that can lead to significant energy and emission savings. This section attempts to identify and delineate such resilient technological solutions that are currently available in the market and can contribute significantly to fuel/energy savings in the transport sector. Therefore, the study determines the amount of emission savings possible through the implementation of these ICT resilient solutions to enable effective use of energy savings. Road transportation is the important mode of transport for movement of both passengers and goods in India, passenger mobility through air and water being inconsiderable in comparison (road, rail and air are responsible for 80 per cent, 13 per cent and 6 per cent, of total transport GHG emissions) [3]. Road transport, one after the other, is completely dependent on fossil fuels, which are the main source of the sector's GHG emissions. Even though, as a long-term strategy, reducing GHG emissions will after some time require decarbonising the transport sector, however, for the short-term, GHG emission reductions can be achieved through the implementation of resilient ICT solutions.

One of the solutions can be enlarging public transport for GHG emission reduction from this sector so that road space can be used more efficiently for the movement of passengers at greater speed. ICT can play a crucial role in enhancing road movement efficiency. While, the largest quantity of GHG emissions reduction cannot be predicted from mitigation measures in the transport sector, however, many of these measures are comparatively low-cost and may be taken on priority on the basis of minimal abatement costs.

Further, such type of technological innovation can be helpful in larger emission reductions at a faster step in spite of changes in travel and settlement patterns. Therefore, deployment of ICT can be supportive for the sustainable development of this sector. In order to achieve reduction of GHG emissions in the transport sector, following three major ICT solutions have been explored:

Mobility Management System

    ★ Intelligent Traffic Management System

    ★ Intelligent public transport system

Supply Chain and Logistic Optimization

    ★ Freight Management System

    ★ Fleet Management System

Telecommuting and Virtual Meetings

    ★ Telecommuting

    ★ Audio conferencing and Videoconferencing [4].

Mobility Management System

Mobility management system includes a wide range of wireless and wired ICT solutions for surface transportation that can significantly enhance the effectiveness and energy efficiency through advanced applications in information, communication and sensors. These systems can be mitigation measure for GHG and congestion reduction and enhance productivity and safety. Management systems like intelligent traffic management systems manage traffic flow through real-time information on incidents, locations, travel times, crowded spots. There are two types of mobility management system.

Intelligent Traffic Management Systems (ITMS)

Intelligent traffic management systems involve smart and advanced tools and appliances for managing transport networks. This system is based on three principle information, analysis and dissemination. These tools offer real –time information about current traffic, which helps authorities, operators and individual travelers to make better informed. Different kinds of technologies such as car navigation; traffic signal control systems; variable messages signs; automatic number plate recognition or speed camera to monitoring applications enhance the efficiency of surface transportation. Further, more advanced tools that integrate live information data and give the guidance and information about parking and weather. These advanced ITMS tools monitors online the day and night traffic at intersections and on the roads on continuous basis through various hardware components. These tools include analytical software which analyze data and video receipt from camera and GPS for exact incident detection and manage the intelligent video surveillance system. 'Online' real-time indicate the traffic volume on roads, accidents and availability of traffic police vehicles on GPS and vehicle tracking systems. Every road junction need different type of light timing set-up and existing systems are working in a distorted sequence. In this situation, an intelligent and efficient traffic lighting control system, with identification of a number of parameters which shows the road conditions, can significantly save energy. Intelligent transport tools can avoid the unpredictable situation at the time of congestion because most of the present intelligent traffic lights are sensor based, that controls the switching operation of the system without any intelligent transport tool.

(i) GHG Emission Reduction

The economic cost of severe congestion is estimated to be as high as 1 per cent to 3 per cent of GDP in developed and developing countries [5]. Fuel economy is related to normal speed, in lack of resilient transport management measure to tackle congestion, there is large gap between official fuel economy rating of vehicles and real achievable fuel economy on the road. Congestion, traffic jam and GHG emission problems can be solved by enhancing traffic capacity at bottlenecks, planning and managing through ICT solutions.

(ii) Better Traffic and Congestion Management Contribute to Decreased Carbon Emissions

Although mobility management initiatives may induce additional traffic to reducing congestion and increasing the carrying capacity, in many circumstances, due to optimization of operating speeds traffic emissions are likely to reduce. It is expected, that traffic management by deploying advanced technology in transportation will reduce GHG emissions as well as improve vehicle efficiency. Studies carried out by different agencies in Delhi show that peak hour speed has dwindled drastically. A Central Road Research Institute (CRRI) study of 2006 shows that during the morning and evening peak hours, 55-60 per cent of major arterial roads in the capital city have travel speeds of less than 30 kmph. Even during off-peak hours, 40-45 per cent of major arterial roads in the capital city have travel speeds less than 30kmph. About 20 per cent or more of the arterial road network is highly congested, with travel speeds falling below 20 kmph throughout the day. A subsequent survey carried out by RITES in 2008 shows that the peak hours peed of 22 kmph is more widespread and that even during off-peak hours the average journey speed is about 26 kmph. This survey also showed that 70 per cent of the road length has peak hour traffic speed of less than 30 kmph. Most vehicles achieve the highest fuel efficiency at speeds above this value [6].

Fuel economy is maximized due to high acceleration and braking. So there is need of fuel- efficient strategy to anticipate traffic conditions, adopt precautionary measures to minimize incidents of acceleration and braking, and maximize coasting time. With the help of advanced ITMS tools accident rates can be decreased, which further reduce the requirement for braking.

(iii) Proportion of Idling Time

Idling causes a severe drop in instantaneous fuel-mileage efficiency to zero miles per gallon. The second largest loss of energy in a vehicle is from idling, or when the engine is on a standby mode. Even at low traffic times, drivers have to wait for longer than is necessary at intersections because the light schedules are designed to serve a large number of vehicles. A self-organizing traffic scheme will eliminate this problem by being responsive to local demands [7].

The CRRI conducted a study, sponsored by the PCRA, on losses of petroleum products at traffic intersections due to idling of vehicles at Delhi, and to recommend remedial measures for conserving fuel. The study classified the intersections into three categories based on the traffic volume and as per this classification, there were 183 high volume intersections, 250 medium volume intersections and 33 low volume intersections. With over 466 intersections having traffic signals, it was seen that 321,432 litres of petrol and 101,312 litres of diesel were being burnt everyday due to idling of vehicles. An ICT driven traffic management scheme for India called 'Developing of Traffic and Communication Network in NCR and Mega Cities and Model System of Traffic Management' has been included for implementation in 11th Five-Year Plan. The scheme has two components, *viz.* (i) Introduction of Intelligent Traffic System (ITS) and (ii) Setting up of an Integrated Data Communication Network (cyber highway). Pushing for induction of IT-driven solutions in traffic management, the Research Committee on Applications of Industrial Electronics of the Union Ministry of Communications and Information Technology (MOCIT) has also suggested setting up of intelligent transport management cells to help mega cities across the country to manage traffic [8].

Although no major region has yet completely been covered by intelligent transport systems, as one of the first initiatives of its kind, an intelligent signaling system (ISS) has become operational on three intersections — at PushpVihar, Pushpa Bhawan and Krishi Vihar — on the pilot Bus Rapid Transit (BRT) corridor in Delhi. KSRTC is also attempting to implement an intelligent transport system project for the city of Mysore, with systems like the vehicle tracking system, real-time passenger information system and central control station [9].

Intelligent Public Transport Systems

Systematic and managed routes and movement of vehicles increase the efficiency and capacity. Deployment of Intelligent Public Transport Systems and other advanced ICT tools can optimize accessibility and efficiency by reducing waiting time, uncertainty, and ensuring safety. GHG emissions can be reduced by greater use of public transport. Fuel consumption in personalized transport is 10 times higher than public transport; therefore, the GHG intensity of travel per passenger kilometer can be reduced through the routing of passengers by public transport. Further, public transport makes the optimum use of road space by carrying the maximum number of people per unit of road space.

Through the use of these advanced tools and technologies the operators and traffic managers can communicate directly to various platforms for dissemination and action, because these tools and technologies provide real-time information regarding the state of traffic and transport network. The vehicle tracking data can also be used to determine journey times and speeds of vehicles through cities, so that traffic demand strategies can be fine-tuned and urban traffic management control can be exercised efficiently to further reduce GHG emissions.

The key ICT solutions, which are the building blocks of this system, are GSM/ GPRS, geographical positioning system, DGPS, dead-reckoning, track circuits, odometers and different combinations of these technologies, electronic display systems and vehicle mounted units, among others. Attempts to deliver GPS-based bus services are underway in Chennai, Ludhiana (private transporters), Delhi (both public and privately managed buses), several cities in Uttar Pradesh (public buses), and Indore (public buses), among others.

Supply Chain and Logistic Management

Supply chain management is the oversight of materials, information, and finances as they move in a process from supplier to manufacturer to wholesaler to retailer to consumer. Supply chain management involves coordinating and integrating these flows both within and among companies. Supply chain includes the planning and management activities like sourcing, procurement, and conversion. Significantly, it encompasses coordination and collaboration with routing partners, which can be suppliers, intermediaries, third-party service providers and customers. Following problems are the major problems of supply chain and logistics management

* Configuration problem in number, location and network missions of suppliers, production facilities, distribution centers, warehouses and customers.

* Distribution strategy problems which include operating control; delivery scheme, mode of transportation(e.g, motor carrier, including truckload, parcel; railroad; ocean freight; airfreight; and transportation control (*e.g.*, owner operated, private carrier, common carrier, contract carrier, or third party logistics

* Problems regarding information which includes the integration of and other processes through the supply chain to share valuable information, including demand signals, forecasts, inventory, transportation and potential collaboration *etc.*

Supply chain and logistics are significant components which make smooth growth of industry and business. In India the logistics industry is evolving rapidly and frequently attempting to reduce costs and provide effective services. In order to improve the performance and efficiency, the industry is relentlessly trying to obtain advanced tools and strategies.

In recent years, manufactures and retailers demand that their supply chain partners conduct the business electronically. Hence, adoption of ICT along the chain has been increasing in a powerful way to integrate the business systems with customers as well as suppliers, laying greater emphasis on the interrelation of such organization by ICT. Due to the focus of manufacturer, retailers and distributors on core business ICT is becoming as significant to shipper as the movement of freight.

There are various critical and technical solutions for supply chain and logistics management and these critical solutions can not only improve the business carbon emission but also enlarge the efficiency of supply chain and logistics.

(i) Logistics Super Grid

Logistics super grid is defined as the global logistic network that integrates logistics service providers and enterprise customers worldwide. This technological model provides advantage to their expertise and improves their efficiency, while supports cost reduction to the beneficiaries.

The concept of a Logistics Super grid and related topics such as Logistics-as-a-Service, Supply Chain on Demand, and Logistics Marketplaces had the potential to become business operating models of the future. The real-time adaptability of a Logistics Super grid would enable flexible collaboration, modular service orchestration, and maximum efficiency at the same time. The transformational approach would keep logistics companies busy for several years to come, as they would have to step up efforts on standardized service modularization and information management, enabling an orchestrated ad-hoc coupling and de-coupling of logistics partners.

(ii) Tracking Systems-Frequency identification (RFID)

RFID is technological tracking system which enables automatic identification and tracking using radio waves. Globally RFID tracking technology is being used by yard management, shipping and freight and distribution centers, especially for tracking containers, pallets, individual products, for car-parking and ticketing. RFID store and retrieve data remotely using markers called RFID transponder. The main advantages of RFID are:

★ It saves time in the operation of input and output material.

★ Improves the tracking

★ RFID is more resistant to its environment like water, mud, shocks.

★ It can read multiple labels and can help reduce the thefts

(iii) GPS- Global Positioning System

A service provider with the help of this receiver can locate and move on land, sea, air or space around the Earth. The GPS system has experienced great success and created a huge commercial development in many areas: shipping, road, and location of trucks. It anticipates delay in deliveries and reduces thefts.

(iv) On Board Computer

On board computer is a device aboard vehicles, that allows to communicate with headquarter, to receive new orders and combined with GPS. It enables communication with the dispatcher and makes possible the transmission of data in real time.

Automatic Guided Vehicles (AGV)

The AGV moves along a pre-established circuit and realize grips and removals of goods in precise places. To move without human intervention, wagons have to know at any time their position. It causes less damage to the goods and reduce the accidents; also it can work in unfavorable conditions.

(v) Warehouse Management System (WMS)

WMS is the integration of various systems, such as computerized vehicle routing, scheduling, packaging and integrated supply chain systems. WMS monitors incoming goods, customer orders and stock levels and this system incorporates mobile transaction recording appliances to record picking and packing activities.

(vi) Fleet Management Systems

Fleet management system is database tool which incorporates the information on different operational areas, such as vehicles, drivers, workshops *etc.* In order to achieve minimum possible idle time this tool manages the necessary administrative and maintenance procedures of vehicles. It is an important tool to help in handling information about and create reports on operational area, like fuel use, accidents and maintenance charges.

(vii) Smartdust

Smartdust is a relatively new development in ubiquitous and economical sensors, and has a promising role to play in ITS. Some of its potential applications include:

* ⭐ Congestion charging: Smartdust can be embedded on the street or road surface and when a commuter's smartdust equipped vehicle passes over the sensor, his account can be charged automatically.

* ⭐ Fleet position reporting: Smartdust enabled vehicles can be tracked when it is within the connected network, permitting the control centre to take intelligent decisions. It can be applied to public transport information systems, or vehicle and freight security.

* ⭐ Parking space occupation and payment: Smartdust embedded in parking spaces can detect if the parking is in use and communicate this information to remote users.

* ⭐ Fast flow tolling: Using smartdust devices in place of conventional microwave transponders, it is possible to enable a fast-flow tolling system.

(viii) Smart Products

Smart products are designed to supply real-time information on the location and condition of goods, equipment and manpower. They are products that can monitor themselves as well as communicate accurate and appropriate information to the relevant party. These products are capable of constant monitoring of parameters such as temperature through wireless sensors and transmitting that data to remote locations.

Telecommuting and Virtual meetings

Teleconferencing is a work arrangement in which an employee's daily commute to a central place of work is replaced by telecommunication links. It is believed to have both positive and negative repercussions for GHG abatement.

Positive impacts of telecommuting on traffic levels and GHG emissions:

★ Reduced travel of employees to work place and associated fuel savings

★ Reduced vehicle ownership as individual telecommuters may not need to own private vehicles

★ Reduced vehicle ownership will also result in increased use of public transport

Effects that may reduce the positive impacts of telecommuting on traffic levels and GHG emissions:

★ Latent demand from people who decide to travel as congestion decreases

★ Leisure travel from telecommuters that take advantage of the commuting time saved increased urban sprawl, facilitated by the diminished need to live in proximity to offices.

However, overall GHG emissions are liable to decline due to reduced travel as a result of telecommuting, irrespective of the negative spin-offs. Virtual meetings are defined as gatherings of two or more people (co-workers) that are mediated by advanced telecommunication devices and do not involve physical contiguity between participants. Virtual meetings can take place through audio conferencing, video-conferencing and more modern forms of telepresence. With the use of virtual meetings, GHGs benefits are derived from avoidance of business related travel, particularly air travel [9].

Role of Resilient Solutions and Information Technology in Transport Planning

Transport sector plays significant role in economic development, however transport planning contains the procedure of planning, funding and physical infrastructural facilities, in addition establishing the institution and organizations to manage and monitor their effectiveness and services. Information and Technology can play important role in this sector. It provides better planning and management with the help of information and tools to enhance resource inputs and outputs. There are different kinds of tools and software and resilient technology available for urban transport sector to improve the quality of information and also reduce congestion and GHG emission. Moreover direct taxation of those that generate congestion tolls can be put in to practice at scales ranging from individual lanes on single links to national road networks.

Resilience of Electricity Sector

Resilience of the energy sector refers to the capacity of the energy system or its components to cope with a hazardous event or trend, responding in ways that maintain their essential function, identity and structure while also maintaining the capacity for adaptation, learning and transformation. Because climate change can create conditions that will negatively impact the energy sector, resilience becomes increasingly important. The resilience "value chain" integrates robustness, resourcefulness and recovery.

★ **Robustness:** the ability of an energy system to withstand extreme weather events as well as gradual changes (*e.g.* sea level rise) and continue operating.

★ **Resourcefulness:** the ability to effectively manage operations during extreme weather events.

★ **Recovery:** the ability to restore operations to desired performance levels following a disruption [10].

Low-Carbon Energy Technology for Energy Sector

To mitigate climate change and increase energy efficiency the energy sector needs low-carbon energy technologies that decrease the fossil fuels use and reduce carbon dioxide. In this context, there is need to explore the GHG mitigation and energy efficiency adaption technologies. There are some examples of energy sector resilience that can be achieved through efficient energy system, renewable energy low-carbon energy technologies.

★ Low-carbon energy technologies such as nuclear and fossil-fuelled plants with Carbon Capture and Storage (CCS) can provide reliable base-load power and reduce the vulnerability of a variable renewable penetration on grid. However, growth in nuclear energy can result in increased cooling water demand, as well as increased public safety concerns with rising extreme-weather risks. Water demand of power plants with CCS can almost double compared to non-CCS plants.

★ More renewable energy can generally reduce water demand compared with thermal power, but new water challenges related to water-intensive concentrating solar power and bio energy production may emerge. Variable renewable energy production is subject to risks of source supply intermittency (*e.g.* wind and solar) that may be exacerbated by climate change. On the other hand, these sources are often associated with a more distributed generation, creating a profusion of electricity sources and a greater ability to localize and buffer disruptions. Expanded regional renewable grid interconnections may present resilience benefits as well as challenges which need to be evaluated on a case-by-case basis. Infrastructure is also subject to risks from high winds, flooding and heavy precipitation. Hydropower can increase pressure on water resources, which may lead to regional water-related conflicts [11].

Grid Modernization

The current electric grid generates power on one end by the use of coal or natural gas, or by harnessing nuclear power and transmits that power for use in homes and businesses on the other end. The process of constantly generating energy and distributing electricity is less efficient than it could be in term of both costs and energy use. After deployment of modern grid it will become easy to enable two-way communication and data flows and operator can better understand the grid's immediate operating status. With the help of this information, operators can run the grid closer to its full potential and capabilities, resulting in greater efficiencies and reliability. That leads to lower costs for utilities and less consumption and lower bills for customers [12].

The Micro Grid Electricity System

A microgrid is a local energy grid with control capability, which means it can disconnect from the traditional grid and operate autonomously. The grid connects homes, businesses and other buildings to central power sources, which allow us to use appliances, heating/cooling systems and electronics. But this interconnectedness means that when part of the grid needs to be repaired, everyone is affected. A micro grid generally operates while connected to the grid, but importantly, it can break off and operate on its own using local energy generation in times of crisis like storms or power outages, or for other reasons. It can be powered by distributed generators, batteries, and/or renewable resources like solar panels. Depending on how it's fueled and how its requirements are managed, it might run indefinitely and it connects to the grid at a point of common coupling that maintains voltage at the same level as the main grid unless there is some sort of problem on the grid or other reason to disconnect. Figure 13.2 shows the microgrid pattern which is the integration of solar power, cell tower, and electrical substation.

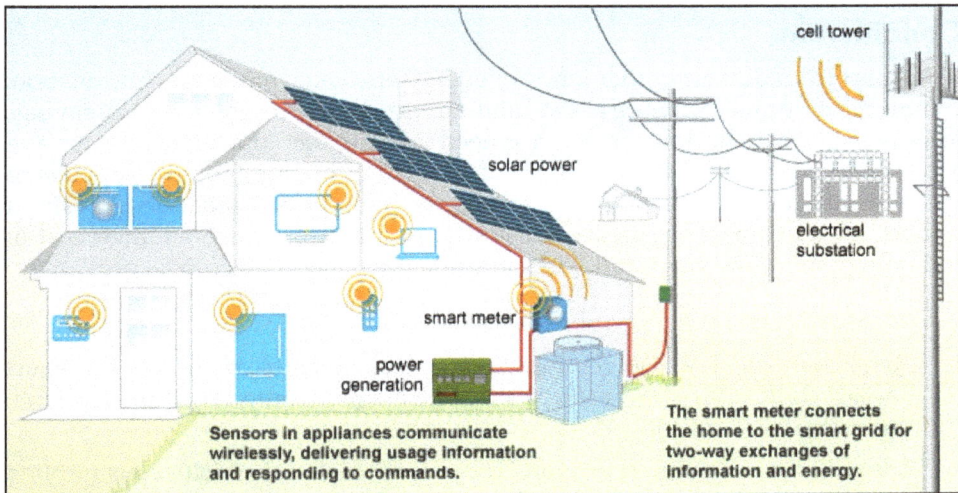

Figure 13.2: Microgrid Pattern at Neighborhood Level (*Source*: www.smartgrid portal. org).

Microgrid is decentralized design that protects consumer from shocks to the wider energy network and enables service in local areas. It increases reliability and maintains hybrid energy supplies from local sources to the grid. At the time of major calamities, microgrid could channel energy to hospitals and other emergency services; it continued to provide electricity, heat, hot water and air conditioning to residents. India's economic growth has led to demand for electricity greatly exceeding supply. The country experiences daily power outages and faces a growing risk of major power cuts [13]. Back-up diesel generators are used to create redundancy and make up shortfalls in supply.

This is an expensive solution, but contributes to severe air pollution in cities [14]. Additional production capacity and modernization of power distribution networks is crucial. *Supervisory Control and Data Acquisition* (SCADA) technology is now being installed, offering the potential to cut power losses in distribution networks by up to 15 per cent [15]. Therefore, deployment of microgrid will help to reduce the demand for local generators, offering economic and health benefits. A Distributed Energy Management System allows these decentralized generation facilities to be operated as a single system, or independently to serve local networks as required. The system helps to regulate variable supplies of power from individual renewable sources and promotes more efficient use of decentralized energy. Flexibility and diversity of supply help to avoid supply disruptions [16]. Energy storage systems like batteries provide additional power at the time of shortage in supply. SIESTORAGE (Siemens Energy Storage) is a modular energy storage system, which uses high performance lithium ion batteries to moderate the output of fluctuating energy supplies. The modular design enables capacity to be adapted to specific demands. The stored electrical energy is used for load regulation (*i.e.* stored power is used when the sun is not shining) and for voltage stabilization [17].

Conclusions

Apart from different other kinds of energy efficiency measures, Information, communication and resilience based solutions can help in reaching energy savings to a significant extent. However, there is need to encourage education, research and accessibility of skilled resources. ICT and resilient technology could be applied to attain the sustainable development goals for policy makers and end users. Advanced technology and smart appliances are costly; in order to reduce the high cost of technology solutions supply and demand side fiscal incentives can be provided.

References

1. Ramachandra, T.V., Shwetmala (2009). Emissions from India's transport sector: Statewise synthesis, Atmospheric Environment, doi:10.1016/j.atmosenv.2009.07.015.

2. Utpal Bahskar (2015). Article on "India's per capita electricity consumption touches 1010 kWh", India. http://www.livemint.com/Industry/jqvJpYRpSNyldcuUlZrqQM/Indias-per-capita-electricity-consumption-touches-1010-kWh.html

3. Ramachandra, T.V., Shwetmala (2009). Emissions from India's transport sector: Statewise synthesis, Atmospheric Environment, doi:10.1016/j.atmosenv.2009.07.015.

4. "ICT's Contribution to India's National Action Plan on Climate Plan", National Mission on Climate Change, A report by CII-ITC Centre of Excellence for Sustainable Development and Digital Energy Solutions Consortium, India.

5. Houghton, J., Reiners, J., Lim, C. (2009). "Intelligent Transport - How cities can improve mobility", USA

6. Central Road Research Institute (CRRI) study of 2006.

7. "ICT's Contribution to India's National Action Plan on Climate Plan", National Mission on Climate Change, A report by CII-ITC Centre of Excellence for Sustainable Development and Digital Energy Solutions Consortium.

8. Central Road Research Institute (CRRI) and PCRA

9. "ICT's Contribution to India's National Action Plan on Climate Plan", National Mission on Climate Change, A report by CII-ITC Centre of Excellence for Sustainable Development and Digital Energy Solutions Consortium.

10. International panel on Climate Change (IPCC) Fifth Assessment Report (AR5) and National Association of Regulatory Utility Commissioners (NARUC). https://www.ipcc.ch/report/ar5/

11. International Energy Agency and International panel on Climate Change (IPCC) Fifth Assessment Report (AR5).

12. Orr, F. (2016). Article on "How Grid Modernization could Improve your life ", USA https://energy.gov/articles/explainer-how-grid-modernization-could-improve-your-life

13. Pidd, H. (2012). Article on " India Blackouts Leave 700 Million without Power", India The Guardian http://www. guardian.co.uk/world/2012/jul/31/india-blackout-electricity-power-cuts.

14. Nessman, R. (2012). Associated Press, "Indian Businesses no stranger to power outage", Ghaziabad, India. http://www.usnews.com/news/world/articles/2012/08/01/power-restored-in-india-but-outage-cause-unclear.

15. Siemens, A.G. (2012). Press Release, "On the road to Smart Grid: Siemens to equip eight cities in India with supervisory control technology for power distribution systems" Nuremberg, Germany. http://www.siemens.com/press/en/pressrelease/?press=/en/pressrelease/2012/infrastructure-cities/smart-grid/icsg201203014.htm.

16. Siemens AG and Stadtwerke München (2012). Press Release, "Stadtwerke München and Siemens jointly start up virtual power plant" http://www.siemens.com/press/en/pressrelease/?press=/en/pressrelease/2012/.

17. Arup, RPA, Siemens (2013). A Research Project on, "Toolkit for Resilient Cities."

18. "ICT's Contribution to India's National Action Plan on Climate Plan", National Mission on Climate Change, A report by CII-ITC Centre of Excellence for Sustainable Development and Digital Energy Solutions Consortium.

19. International Energy Agency.

20. www.smartgrid portal.org.

Chapter 14

Plastic Tank Biogas Digester: A Case Study with Energy and Cost Analysis

W.A.L. Sunil Karunawardana

Engineer,
National Engineering Research and Development Center (NERDC),
Ekala Ja Ela, Sri Lanka
E-mail: karunawardanasn81@gmail.com

Abstract

The accumulation of garbage in urban areas from the households has become a major problem for the municipal, urban and town councils of this country. To give a solution for this in a sustainable way, a domestic bio gas unit was introduced. But there are some problems in those units like clogging and low efficiency.

In this research it is aimed to increase the efficiency of bio gas unit and to overcome the clogging. The proposed bio gas unit was fabricated and tested for kitchen waste. The test results highlighted that a higher amount of bio gas could be obtained and clogging could be overcome. The outcome of this research can be applied for houses located in the municipalities and urban areas, managing the food waste within the home premises. The advantage of this biogas unit is to provide a solution for domestic garbage disposal replacing a part of the cost for LP gas, and provide a good liquid fertilizer for home cultivation. Further it contributes to reduce the climate change by reducing the greenhouse gas CH_4 released to the environment.

Keywords: Biogas, Garbage, Anaerobic digestion, Food waste, Sustainability.

Potential Problems Associated with Food Waste

Improper disposal of cooked and uncooked food is a health and environmental issue. Every kitchen in houses, restaurants, hotels, hospitals, hostels *etc.* throw cooked and uncooked food waste daily. These food wastes become rotten or decomposed and emit bad smell while generating hazardous bacteria.

Also such rotting food waste is a good breeding ground for flies. Flies can propagate food borne diseases. Therefore it is essential to have a proper food waste disposal/treatment practice.

Gravity of Food Waste Issue in Urban Life – The Reflection of Garbate Problem in Colombo

Daily food waste generation is an issue, especially in urban areas. Most of the individual houses in Colombo have no self-sustaining individual arrangement of garbage disposal. In urban areas daily garbage collection by municipalities is the prevailing practice.

Daily Food and Bio-degradable Waste Accumulated in Colombo

The data given below shows that garbage generation in Colombo is an indication of the grim nature of garbage management in modern day living.

The Places/Sources of Garbage Generation

Houses, hotels, restaurants, hospitals, offices and other establishments

Daily Garbage Collection

Number of houses in Colombo municipality area: *Approx. 473,045* [4]

Daily garbage collection (if considered 0.75 kg per day from each house): *Approx. 473045×0.75 = 355 tons*

Number of tourist/regular hotels and restaurants in Colombo: *Approx. 2500*

If daily average garbage collection from each hotel is *approx. 25kg,*

Daily garbage collection from hotels and restaurants: *Approx. 2500 ×25 kg = 62.5 tons*

From other institutions like hospitals and offices: *Approx. 12.5 tons*

Total garbage from all these places: *Approx. (355 + 62.5 + 12.5) = 430 tons*

(These are the estimated value according to the census of population and housing 2001 [4])

The total daily solid waste collection in Colombo is 700 tons and 430 tons of food and bio degradable waste form about 60 per cent of the daily solid waste generation [1].

Dumping – The Existing Practice with Solid Waste in Colombo

Dumping is the current practice of managing garbage in Colombo city. Looking for new dumping ground and the community uprising against dumping are always a topic for the media. About 30,000 m³ of methane (CH_4) emit from these garbage dumps to the environment daily. Methane release is one of the causes for damaging the ozone layer.

Also this primitive level practice of garbage dumping damages the beauty of the environment. Dumping causes health issues, inconvenience, and strong bad smell. There is a continuous struggle between the community and the local government due to the dumping issue. Also there is an issue of finding sites for dumping; transporting rotten bio waste through the city is disgusting.

Small Scale Solution for the Garbate Problem – Bio-digester

Why is Garbage a Problem?

Having taken the responsibility and the ownership of daily garbage collection in the city, Colombo municipality is facing lot of hardships and humiliations. Is this garbage ownership is because of the tax being collected from the households? Why is the municipality continuously failing on garbage issue? Why did garbage become a problem, instead of an answer?

Giving the Garbage Ownership Back to the Origin (households) with Technological Interventions – The Right Approach

It is a fact that the municipalities taking the ownership of garbage is a problem over decades. Problem seems further escalating, rather than diminishing. Hence, it would be wise to give the garbage ownership back to households, hotels, and other institutions, where the garbage is originated. At the same time the authorities like municipalities and environmental agencies should intervene with technologies, to let the households themselves handle the garbage. Bio-digester would be the best intervention for such decentralized approach.

Biogas Digester for Garbage and Food Waste in Households

Food waste (C, H, O, S, N) when decomposed through anaerobic digestion produces methane which can be used as a heat energy source for cooking, and the effluent from the biogas digester can be used as liquid organic fertilizer. There would be no more solid waste since liquid organic fertilizer can be used for kitchen gardening. This helps the household to reduce spending on cooking gas and in minimizing the use of inorganic fertilizer. Within Colombo, to individually manage at least half (215 tons) of 430 tons, each and every house or dwelling unit is to be equipped with Plastic bio gas tank of relevant capacity according to the daily garbage collection.

What is Anaerobic Digestion?

Anaerobic digestion is a collection of processes by which microorganisms break down biodegradable material in the absence of oxygen. The process is used for industrial or domestic purposes to manage waste and/or to produce fuels.

The Mega Scale Centralized Bio-digester against the Domestic Scale Bio-digester

Centralized bio digester that operates by collecting household food waste to handle over 430 tons per day involves heavy labor, infrastructure, and resourceful system of operation. The investment as well as operation involves a large amount

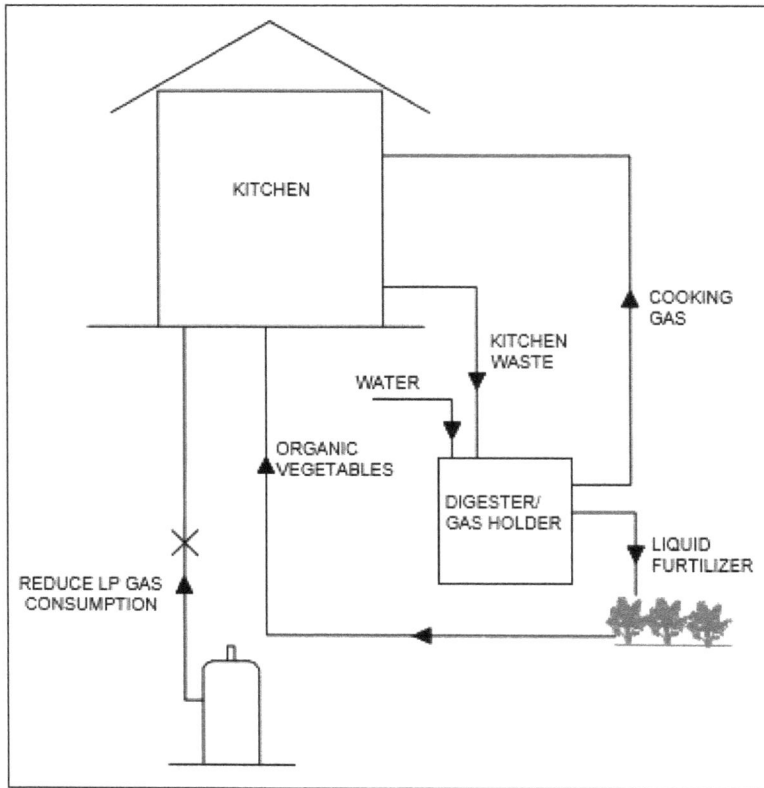

Figure 14.1: Kitchen Waste Disposing System.

of effort and money. If these 430 tons of garbage is to be treated with no side issues, mega investment is needed, to the tune of Rs.4880 million the decentralized domestic scale bio digester makes every household responsible for their food waste disposal.

Thus the practice of keeping the responsibilities of waste management with those who are responsible for waste generation could be established. In the year 2000, there were 50 million household bio gas digesters in China.

Plastic Tank-Digester for Household Food Waste: A Case Study

A Plastic tank which has the capacity of 1000 litres was used for this study and it was modified as a continuous type bio digester. As the gas volume is to be measured, separate gas holder of 175 litre capacity was used to store gas. Initially cow dung was fed into the digester once in every three days over a period of two weeks, that is 40, 40, 35, and 20 kg respectively and made the bacteria culture before feeding food waste.

Gas samples were analyzed during this period using gas chromatograph and obtained the percentage of methane and other gases. Here 25 mm height of gas

Figure 14.2: The Plastic Digester with Gas Holder.

Figure 14.3: Side View of the Digester.

holder depicts 5.5 liters of gas (the total volume of gas holder is 170 liters and the total height is 775mm.)

Findings

On feeding only the food waste (kitchen waste) to the digester, it was observed that 156 litres of bio gas could be obtained from 1 kg of food waste over 24 hours under normal atmospheric pressure and temperature.

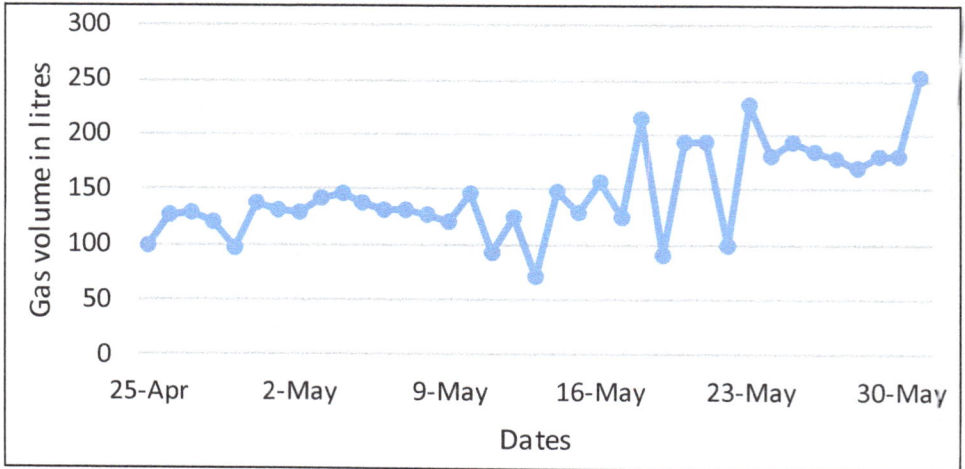

Figure 14.4: Generation of Gas for 35 kg Kitchen Waste during 35 Days. The daily feed varied from 600 g to 6 kg. It depended on the food waste generation on each day in the kitchen.

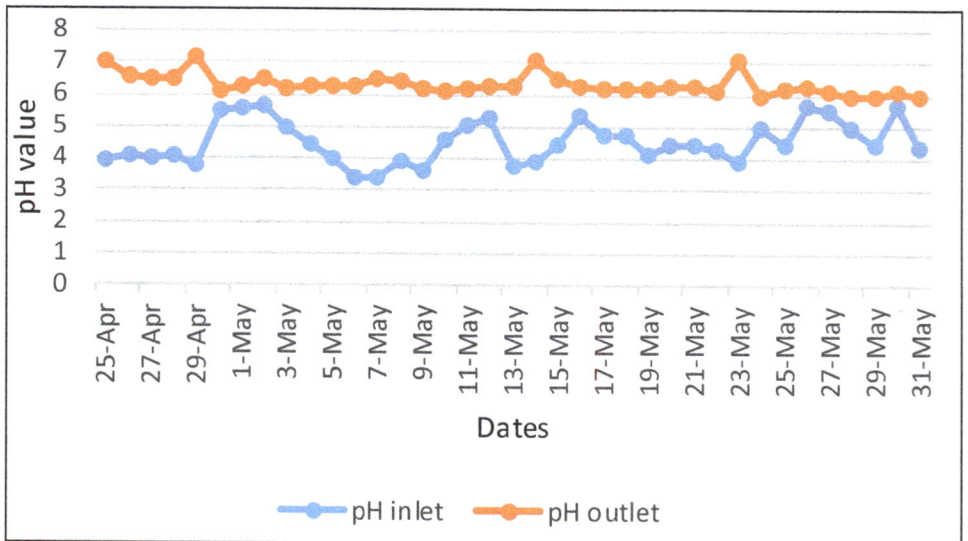

Figure 14.5: pH Variation when Feeding Food Waste for 35 Days.

Methane generation in this bio gas system remains 40 to 60 per cent. The pH variation when feeding only food waste was 4 to 6 at inlet and 6 to 7.2 at outlet. The slurry which comes out from the digester was tested for Chemical Oxygen Demand and Biological Oxygen Demand, and the values are:

COD =3540 mg/litre, and BOD =930 mg/litre.

Average bio gas generation is 157 litres from 1kg per day

Table 14.1: Cost Comparison of Biogas with LPG

Weight of Bio gas of 1 m³	0.9 kg
Cost of 1 kg of LPG = 1500.00/12.5	Rs120.00
Average bio gas from 1 kg of food waste	157 litres
Then the total gas collection for a month	157 x 30 = 4710 litres.
Bio gas generation for a month	4.7x 0.9 = 4.2 kg
This gas amount is approximately 1/4 of a normal LPG cylinder (12.5 kg)	
Cooking time for 157 litres of gas	31 minutes
The total cooking time per month by using 1 kg food waste per day is 15 hours and 30 minutes	
Cost of 1 kg bio gas	Rs 100.00
Then the value of gas	4.2 x 100 = Rs 420.00

Table 14.2: Biogas Volume Needed for Cooking some Food Items (Observed Values)

Sl.No.	Food Item	Food Weight	Time Taken (minutes)	Gas Volume in Liters
1	Rice (samba)	1kg	30	150.00
2	Milk rice	1kg	35	165
3	Dhal	200g	21	93.5
4	Fish	300g	24	110
5	Bean	250g	21	99
6	Potato	250g	24	93.5
7	Ash Banana	250g	25	115.5
8	Brinjal	250g	20	99
9	Ladies' fingers	250g	15	60.5
10	Pumpkin	500g	22	104.5
11	Canned fish	250g	18	82.5

Conclusions

After analyzing the data it could be concluded that, food waste of 1 kg gives about 150 litres of gas per day.

The main advantage of this system is that it helps to minimize the garbage accumulation to the road sides of urban areas and stops emission of greenhouse gas, CH_4.

Table 14.3: Comparison of 1000 Litres Plastic Tank with 6m³ Conventional Biogas Unit (Values tested at NERDC lab)

Sl.No.	Parameters	Conventional Biogas System (6 m³ continuous type digester)	1000 Litres Plastic Tank Biogas System
1	Material of Construction	Bricks, Sand, Cement, Concrete and Steel	1000 litre plastic tank and two barrels
2	Space required	Large	Small
3	Quantity of feed stock	Nearly 45kg + 45litres water per day for 6 m³ digester	1 to 4 kg.Waste + 3 to 6litres of water daily
4	Type of feed stock	Animal dung/market garbage	Food waste
5	Quantity and nature of slurry to be disposed	About 40 to 50 (semi solid slurry)	About 5-6 litres of liquid leachate
6	Reaction time for full utilization of feed stock	20 days	12 hours
7	Standard size to be installed	6,000.00 litres	1000 litres
8	Production of bio gas	1 m³ for 25 kg food waste	150 litres from 1 kg
9	Methane in bio gas	40 to 60	40 to 60
10	Calorific value	22.7 MJ/m³	22.7 MJ/m³ (5)
11	BOD of slurry	642 mg/L (Market Garbage)	930 mg/L
12	Maintenance	high	low
13	COD	9192 mg/L (M/G)	3540 mg/L
14	Cost	About Rs175,000.00	About Rs45,000.00

There are two other advantages of this unit in addition to the bio gas. It gives a good organic liquid fertilizer and this is a solution to dispose garbage in a useful way. This Plastic digester is a new design and it has been designed to minimize the clogging of feeding materials.

This will be a good solution for the domestic garbage problem in Sri Lanka.

Acknowledgements

The author is grateful to Dr Warahena for technical advise to prepare this research paper, to Mr. Parakrama Jayasinghe, former member of NERDC director board for encouragement to do this research And to NERDC for affording resources to conduct the research.

References

1. http://www.statistics.gov.lk/pophousat/cph2011/pages/activities/reports/cph_2012_5per_rpt. pdf

2. https://newscenter.nmsu.edu/articles/view/5379

3. http://www.ijsrp.org/research-paper-1113/ijsrp-p2357.pdf

4. Census of population and housing, 2001

5. The bio gas hand book, Edited by Arthur Wellinger, Jerry Murphy and David Baxter

Chapter 15

ANN and PI Based Fuzzy Logic Controller for Wind Driven Self Excited Induction Generators

Hussein F. Soliman[1], Abdel-Fattah Attia[2]
M. Mokhymar Sabry[3] and M.A.L.Badr[4]*

[1,4]**Faried Consult Office, Electric Power and Machine Department,*
Faculty of Engineering Ain Shams University,
E-mail: faried_office@yahoo.com
[2]*Head of Electrical Engineering, National Research Institute of Astronomy and Geophysics*
"NRIAG", Helwan, Cairo, Egypt
[3]*Electricity and Energy Ministry, New and Renewable Energy Authority*
"NREA" Wind Management
E-mail: sabry0001@gmail.com

Abstract

Enhancement of performance of Self-Excited Induction Generator (SEIG) driven by Wind Turbinehas been addressed using two types of reactive and active power controllers in this paper. Also a comparison has been made between the two controllers with PI_FLC and ANN_FLC., PI_FLC is tuning its variable integral gain using Fuzzy Logic Controller,andANN_FLC has a variable learning rate which is adapted by using FLC. The SEIG system is equipped with two controllers to regulate the terminal voltage by adjusting self excitation, *i.e.* reactive power control, and regulate the mechanical power or stator frequency by regulating the blade angle of WT, *i.e.* active power control. Different simulation results are obtained while the system is subjected to sudden disturbance in the isolated load. PI_FLC and ANN_FLC performance characteristics are discussed for comparison.

Keywords: Induction, Generator, Controllers, FLC, ANN, PI, Wind.

Introductrion

Studies of Self-Excited Induction Generators (SEIG) have been on since 1935. Many researchesin the field of SEIG have been targeting to overcome different problems. The primary advantages of SEIG are low maintenance cost, better transient performance, lack of dc power supply for field excitation, brushless construction (squirrel-cage rotor), *etc.* In addition, the induction generators have been widely employed to operate as wind-turbine generators and small hydroelectric generators in power systems. The induction generators inject electric power to large power systems when the rotor speed of the induction generator is greater than the synchronous speed of the air-gap-revolving field.

This research concentrates on the dynamic performance of an isolated SEIG, driven by Wind Turbine (WT), to supply an isolated static load. As known, the d-q axes equivalent circuit model based on different reference frames extracted from fundamental machine theory can be employed to analyze machine's transient's response in dynamic performance. However, the previous investigators have rarely studied the SEIG controlled by the artificial intelligence (AI) based controllers, to supply isolated loads. This paper studies the performance of the SEIG, equipped with the switching capacitor bank, using one AI controller to adjust the duty cycle. The major contribution of this research is the application of ANN in the controller with on-line adaptation for the weights and neuron biasing through the BP algorithm. Also, the effects of changing some parameters such as the learning rate for ANN and Integral gain of PI using FLC are conducted. This paper combines the linearized saturation curve of magnetizing reactance (X_m) versus magnetizing current (I_m) and machine's d-q axes dynamic model.

System Under Study

The mathematical model of SEIG driven by WT is simulated using MATLAB/Simulink package to solve the differential equations. Meanwhile, two controllers have been developed for the system under study. The first one is the reactive power controller to adjust the terminal voltage at the rated value, through the variation of the self-excitation using switching capacitor bank, by a controlled duty cycle. The second controller is the active power controller to control the input mechanical power to the generator and thus maintain the stator frequency constant by changing the value of the blade pitch angle of the wind turbine. Both controllers use two types of tools;the first is the Proportional Integral hybridised with Fuzzy Logic Controller (PI_FLC) to adapt a variable integral gain (K_{IV}). The second adapts the learning rate (α_r) of the ANN using the fuzzy set. The simulation results depict the variation of the different variables of the system under study, such as terminal voltage andfrequency. Figure 15.1 shows the system under study, which consists of SEIG driven by WT and connected to isolated load;there are two control loops for terminal voltage and pitch angle using the PI_FLCand ANN_FLC.

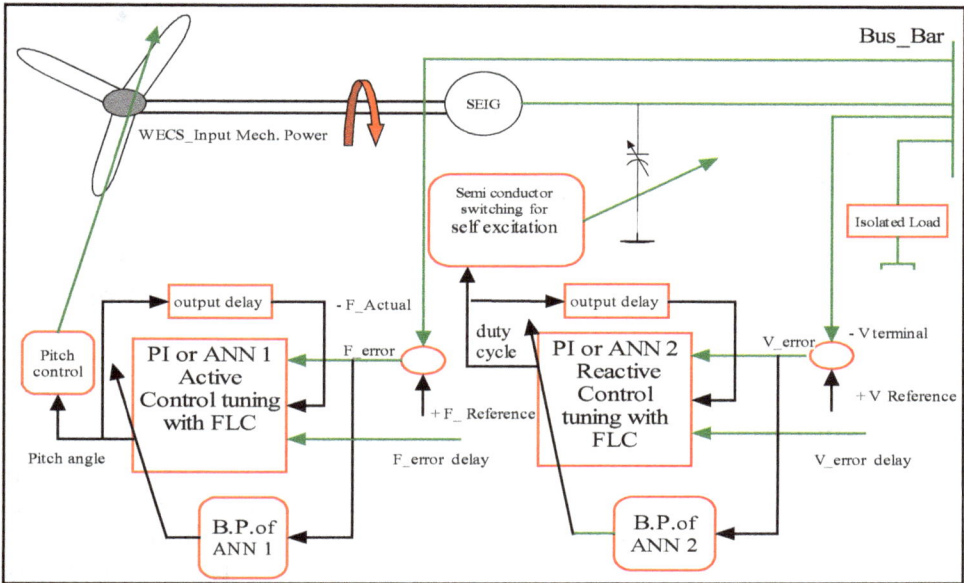

Figure 15.1: System under Study.

Mathematical Model of Seig System and its Control

Electrical System

Figure 15.2a-b shows the d-q axis equivalent-circuit model for a three-phase symmetrical induction generatoron blocked rotor,. The stator and rotor voltage

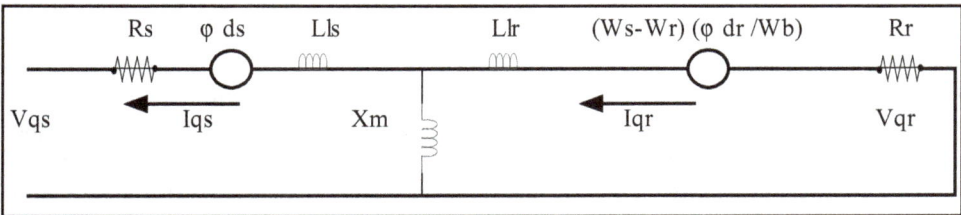

Figure 15.2a: Equivalent Circuit of an Induction Generator for Direct Axis.

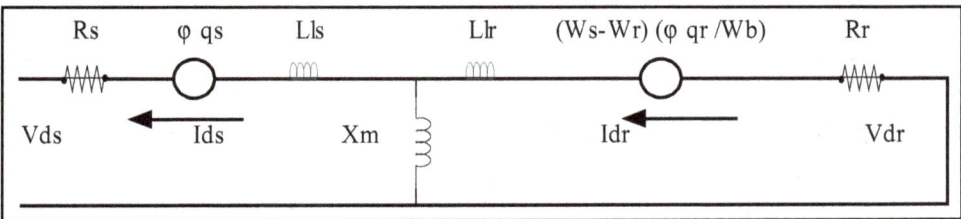

Figure 15.2b: Equivalent Circuit of an Induction Generator for Quadrate Axis.

equations using Krause transformation [1,2,3], based on stationary reference frame are given in Appendix A.

Wind Turbine

The equation between the power coefficient of the wind turbine, tip speed ratio (μ) and pitch angle (β) is given in Appendix as (A5). The analysis of SEIG in this research is based on the following assumptions [3,4]:

★ All parameters of the machine can be considered constant except X_m.

★ Per-unit values of both stator and rotor leakage reactances are equal

★ Core loss in the excitation branch is neglected.

★ Space and time harmonic effects are ignored.

Equivalent Circuit

The d-q axes equivalent-circuit models for a three-phase symmetrical induction generator on blocked rotor, are shown in Figures 15.2a and 156.2b. The equivalent-circuit parameters shown are for a (1.1 kW, 127/220 V (line voltage), 8.3/4.8 A (line current), 60 Hz, 2 poles), wound-rotor induction machine [1,2,3]. By choosing proper base values: base voltage (V_b)= [220/(1.73)] V, base current (I_b)= 4.8 A, base impedance (Z_b)= 26.462 ohm, base rotor speed (N_b)= 3600 rpm, and base frequency (F_b)= 60 Hz, the per-unit parameters of the induction machine under study are equal to: stator resistance (R_s) = 0.0779, rotor resistance (R_r)= 0.0781, stator reactance (X_s) and rotor reactance (X_r) are equal 0.0895. The equation of motion of rotating part of the SEIG coupled withthe wind turbine is also included in the system for having a detailed simulation model. The inertia constant of the machine (H)= 0.055 s.

Reactive Control and Switching Capacitor Bank

Switching

The switching of capacitors has been discarded in the past because of the practical difficulties involved [5,6], *i.e.* the occurrence of voltage and current transients. It has been argued, and justly so, that current 'spikes' for example, would inevitably exceed the maximum current rating as well as the (di/dt) value of a particular semiconductor switch. The only way out of this constraintwould be to design the semiconductor switch to withstand the transient value at the switching instant. An equivalent circuit in Figure 15.3 is added to explain this situation of switching capacitor bank and the duty cycles. For the circuit of Figure 15.3, the switches are operated in anti-phases, *i.e.* the switching function f_{s2} which controls switch S_2 is the inverse function of f_{s1} which controls switch S_1. In other words, switch S_2 is closed during the time when switch S_2 is open and vice versa.

Reactive Control through Switching Capacitor Bank Technique

In the system under study in Figure 15.1, the input to the controllers is the voltage error and the output of the controllers perform the duty cycle (λ) which is used to compute the effective capacitor bank value (C_{eff}). The duty cycle (λ) or the output of the controllers is used as an input to semiconductor switches to change

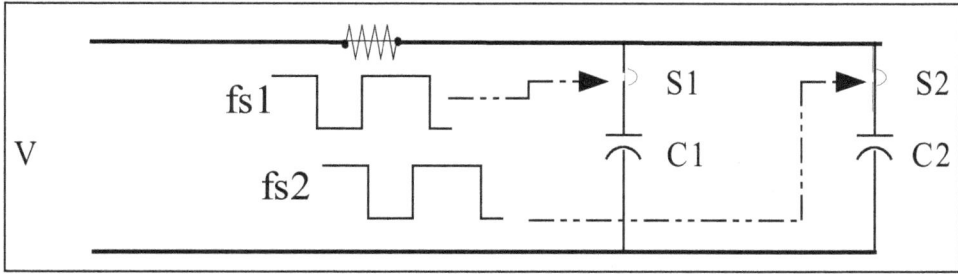

Figure 15.3: Semi Conductor Switches (S$_1$,S$_2$) for Capacitor Bank.

the capacitor bank value as per the need of the excitation and according to the controllers output. Thus the terminal voltage is controlled by adjusting the self-excitation through automatic switching ofthe capacitor bank.

Active Control through Variable Pitch Control Technique

Active control is applied to the system under study by adjusting the pitch angle ofwind turbine blades. This is used to maintain the SEIG at a constant stator frequency and reject the effect of the disturbance. The pitch angle is a function of the power coefficient, C_p, of the wind turbine. The value of C_p is calculated with the pitch angle according to equation (A5). Consequently, the best adjustment for the value of pitch angle leads to improve the mechanical power regulation, which achieves a better adaptation of system frequency. Accordingly, the active power control regulates the mechanical power of the wind turbine.

Controllers

Two types of controller are studied; the first is the Proportional plus Integral controller (PI) tuning its integral gain with FLC but the Proportionalgain (K_p)is kept constant. The second is ANN controller tuning its learning rate by FLC.

PI-Controller Technique

The Proportional plus Integral (PI) controller is a conventional controller which has a variable Integral gain (K_{IV}), and constant Proportional gain (K_p). Forthe best stetting of gains the best control could be obtained. The voltage or frequency error issubjected as an input tothe PI controller, then the output is supposed as the duty cycle for the switching capacitor bank in the reactive controller;but in the active controller the output is supposed to tune the pitch anglevalue of the wind turbine. The technique of having variable K_{IV} depending on the voltage error, for voltage control, is introduced to obtain the advantage of high and low valuesof the integral gain of voltage loop. A program is developed to compute the value of the variable integral gain K_{IV} using the following rule base:

if $(e_V < e_{V\,min})$,

$\quad K_{IV} = K_{IV\,min}$;

elseif $(e_V > e_{V\,max})$,

$\qquad K_{IV} = K_{IV\,max}$;

else $(e_{V\,min} < e_V < e_{V\,max})$,

$\qquad M = (K_{IV\,max} - K_{IV\,min})/(e_{V\,max} - e_{Vmin})$;

$\qquad C = K_{IV\,min} - M \times e_{V\,min}$;

$\qquad K_{IV} = M \times e_V + C$;

end

where,

e_V = the voltage error, e_{Vmin} = the minimum value of the voltage error, e_{Vmax} = the maximum value of the voltage error, K_{IVmin} is the minimum value of K_{IV}, K_{IVmax} is the maximum value of K_{IV}, C is a constant and M is the slop value. Figure 15.4 shows to the plot of K_{IV} calculated using the rule base, against the terminal voltage error e_V. The values of the e_{Vmin} and e_{Vmax} are obtained by trail and error to give the best dynamic performance [3].

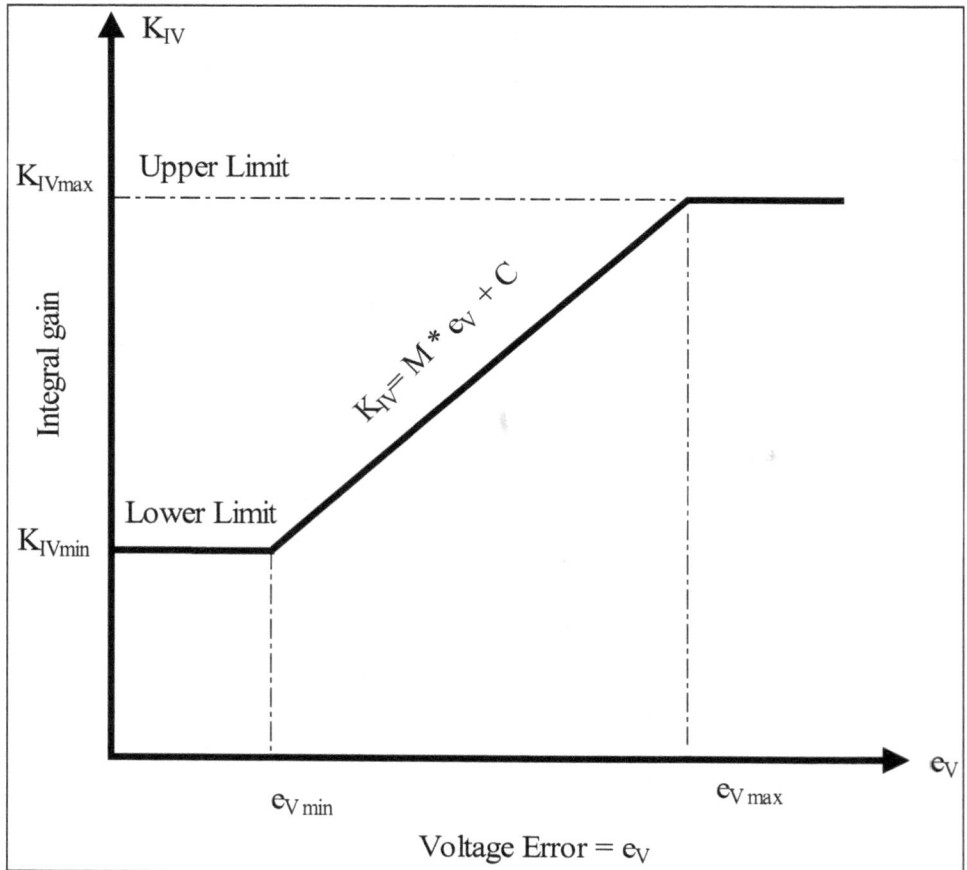

Figure 15.4: Variable Integral Gain for PI Controller.

ANN Controller Technique

The architecture of the ANN is depending on the construction of the biological neuron. The proposed ANN consists of three layers; namely, input, hidden, and output layer. The input layer has three nodes. The input vector to the ANN consists of the terminal voltage error, the terminal voltage error at previous instant (voltage error {k-1}), and the output of the ANN at previous instant (output {k-1}). The output layer has a single neuron, while the hidden layer has two or five neurons. Both hidden and output layer neuron activation function is "Logistic- Function" [7,8,9,10].The mathematical equations, describing the relations of the ANN, are given in Appendix A as (A6) and (A7). There are various techniques to update the connection weights and neuron bias has been used. The delta rule (error adaptation) using the Back Propagation technique is used in the present research. The main parameter in the weight adaptation is the learning rate α_r, which could be determined upon the system situations. The used ANN had three input variables in the input layer, one hidden layer with two neurons, and finally the output layer consists of one neuron. This ANN could be so-called (3-2-1 ANN, means 3 inputs, 2 neurons in the hidden layer and 1 neuron in the output layer). Figure 15.5a shows the schematic block diagram of (3-2-1) ANN controller. In this task the effect of changing learning rate α_r is studied.

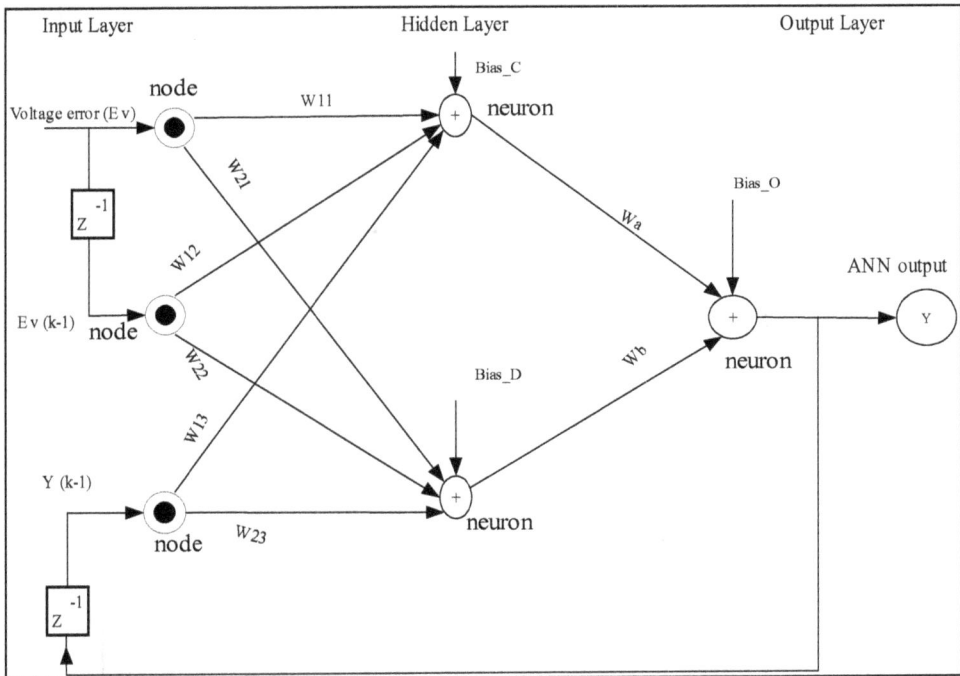

Figure 15.5a: ANN - Construction for Three Input Variables; Two Neurons in Hidden Layer, One Neuron in the Output Layer.

As explaind above a program is developed to calculate the variable K_{IV} for PI; the same technique is applied to adapt the variable learning rates (α_{rv}, α_{rf}) for reactive and active power controls respectively; using the following rules base, for example in reactive power control:

if $(e_V < e_{Vmin})$,

$\alpha_{rv} = \alpha_{rv\,min}$;

elseif $(e_V > e_{Vmax})$,

$\alpha_{rv} = \alpha_{rvmax}$;

else $(e_{V\,min} < e_V < e_{V\,max})$,

$\qquad M = (\alpha_{rv\,max} - \alpha_{rv\,min}) / (e_{V\,max} - e_{V\,min})$;

$\qquad C = \alpha_{rv\,min} - M \times (e_{V\,min})$;

$\alpha_{rv} = M \times (e_V) + C$;

end;

where,

e_V = the voltage error, e_{Vmin} = the minimum value of the voltage error, e_{Vmax} = the maximum value of the voltage error, α_{rvmin} is the minimum value of α_{rv}, $\alpha_{rv\,max}$ is the maximum value of α_{rv}, C is a constant and M is the slop value. Figure 15.5b

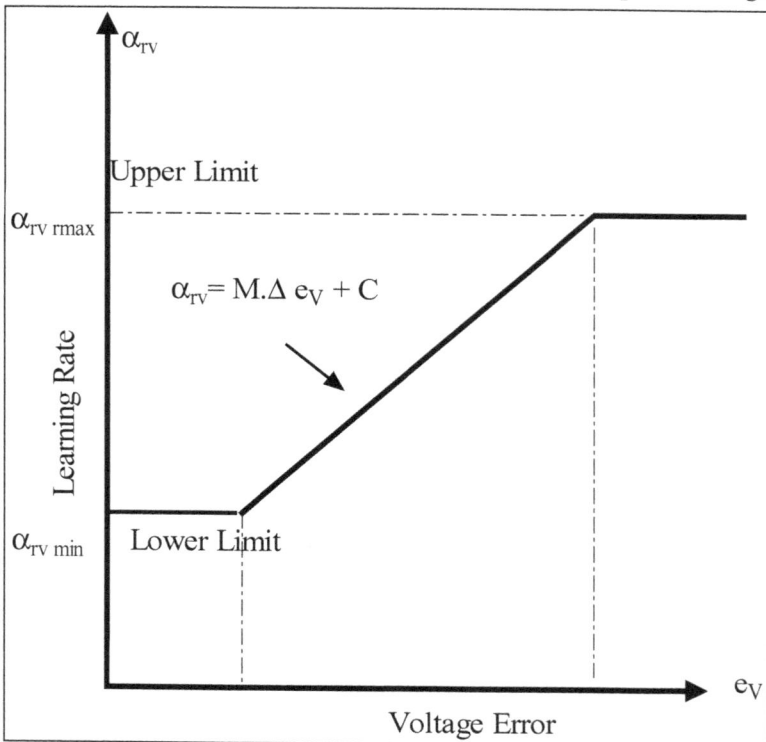

Figure 15.5b: Variable Learning Rate between Upper and Lower Values.

shows α_{rv} calculated based on these rulesand plotted against the terminal voltage error e_v. The value of the e_{Vmin} and e_{Vmax}areobtained by trail and error to give the best dynamic performance.

Fuzzy Logic Controller (FLC)

To design the FLC, the control engineer must gather information on how the artificial decision maker should act in the closed-loop system, and this would be done from the knowledge base [11,12,13]. Fuzzy system is constructed from input fuzzy sets, fuzzy rules and output fuzzy sets, based on the prior knowledge base of the system. Figure 15.5c shows the basic construction of the FLC. There are rules to govern and execute the relations between inputs and outputs ofthe system. Every input and output parameter has a membership function which could be introduced between the limits of these parameters through a universe of discourse. The better is the adaptation of fuzzy set parameters the better tuning of the fuzzy output is obtained. The proposed FLC is used to compute and adapt the variable integral gain K_{IV} of PI controller and the variable learning rates (α_{rv}, α_{rf}) of ANN controller.

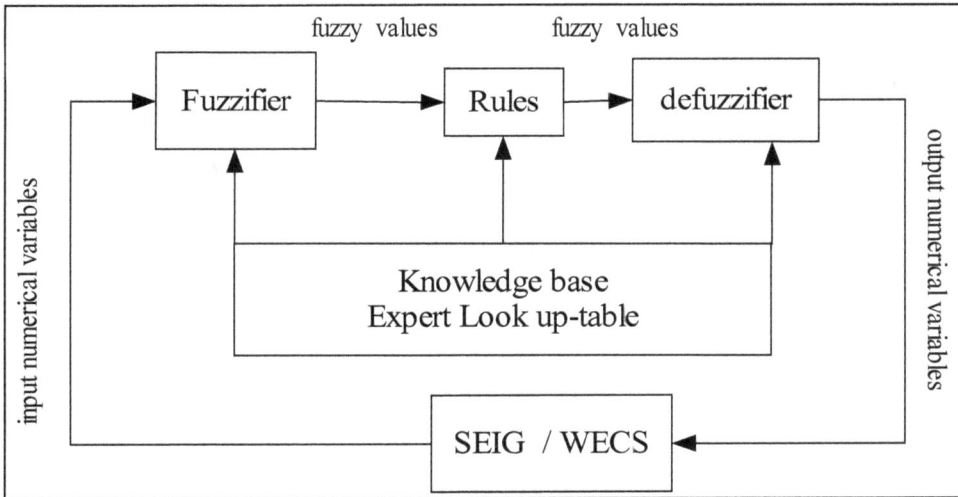

Figure 15.5c: The Three Stages of Fuzzy Logic Controller.

Results and Discussion

Simulation Results

Based on the mathematical model of the system under study, equipped with two controllers (PI_FLC and ANN_FLC) for terminal voltage and stator frequency, the simulation is carried out using the MATLAB- Simulink Package.

PI_FLC Simulation Results

The simulation program is carried out for different values of K_{IV} while the value of the proportional gain is kept constant. It is noticed from the simulation results that the value of percentage overshoot, rising time and settling time change as K_{IV} is changed as in Figure 15.6. Running for PI controller with various integral gainsneedsa relation between the voltage or frequency error and the value of this gains.These gains are computed here using FLC instead of Matlab. Figures 15.7(a) and (b), and Figure 15.8 show the membership function of error, change of error and output respectively. Table 15.1 shows the lookup table of fuzzy set rules for reactive power control. Figure 15.9 shows the comparison between terminal voltage versus time for PI performance with and without fuzzy set technique to compute the variable integral gain. Figure 15.10 shows the comparison between stator frequency versus time for PI performance with and without fuzzy set technique.

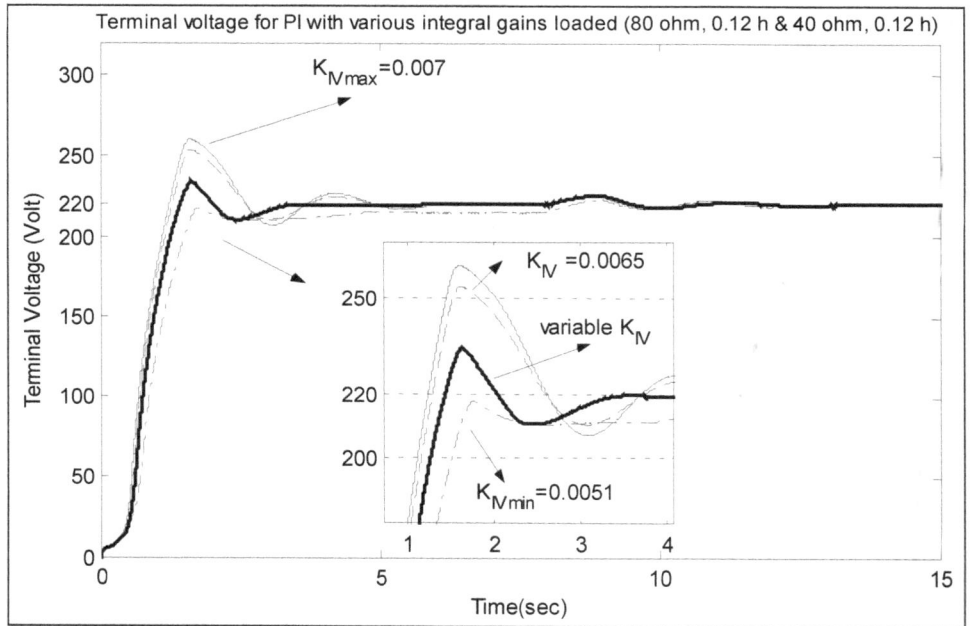

Figure 15.6: Dynamic Response of the Terminal Voltage with different Values of Integral Gain for Voltage Control.

Table 15.1: Fuzzy Set Rules for Reactive Power Control

Ve_ error	Ve'_ change of error		
	LO	AV	HI
LO	LO	LO	LO
AV	AV	AV	AV
HI	HI	HI	HI

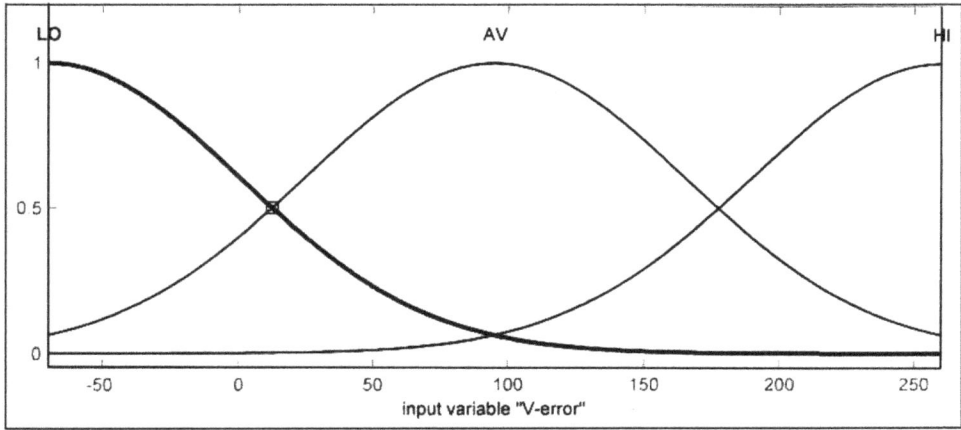

Figure 15.7a: Membership Function of Voltage Error.

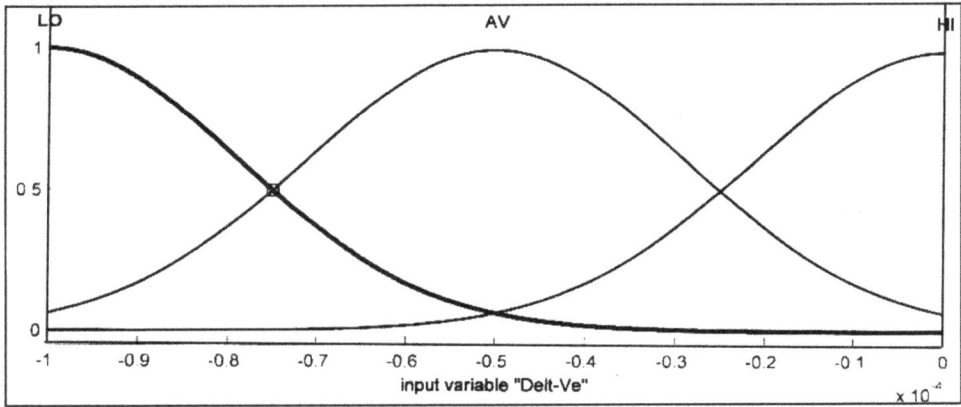

Figure 15.7b: Membership Function of Change on Voltage Error.

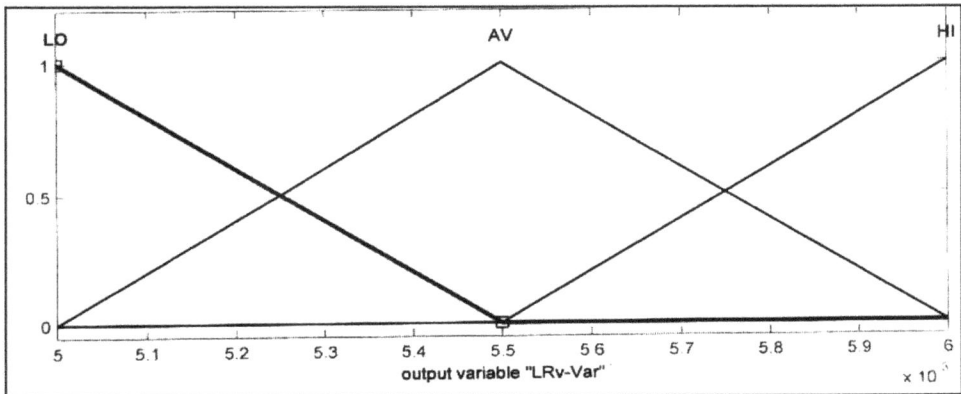

Figure 15.8: Membership Function of Variable Integral Gain.

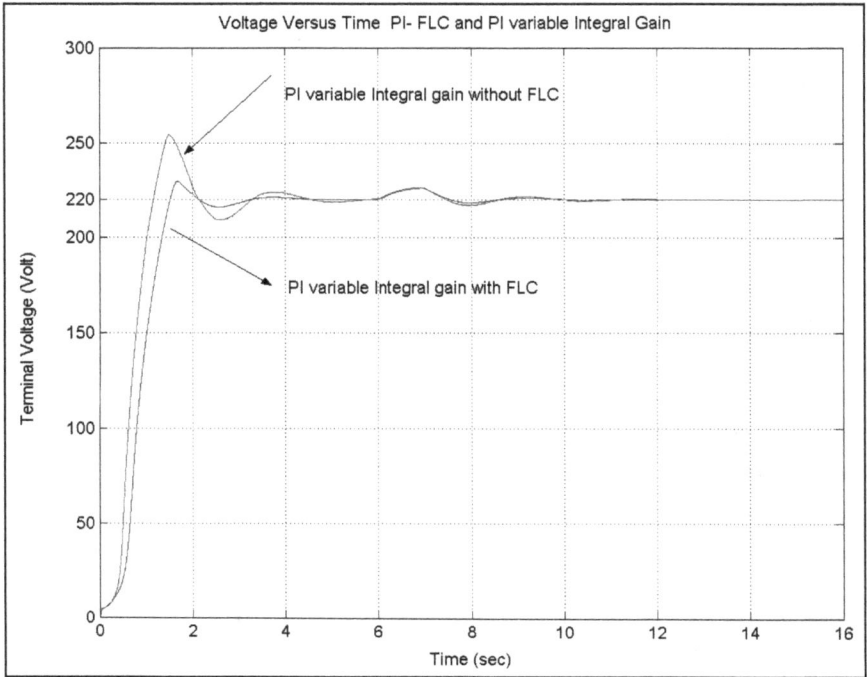

Figure 15.9: Dynamic Response of Terminal Voltage for PI with and Without FLC.

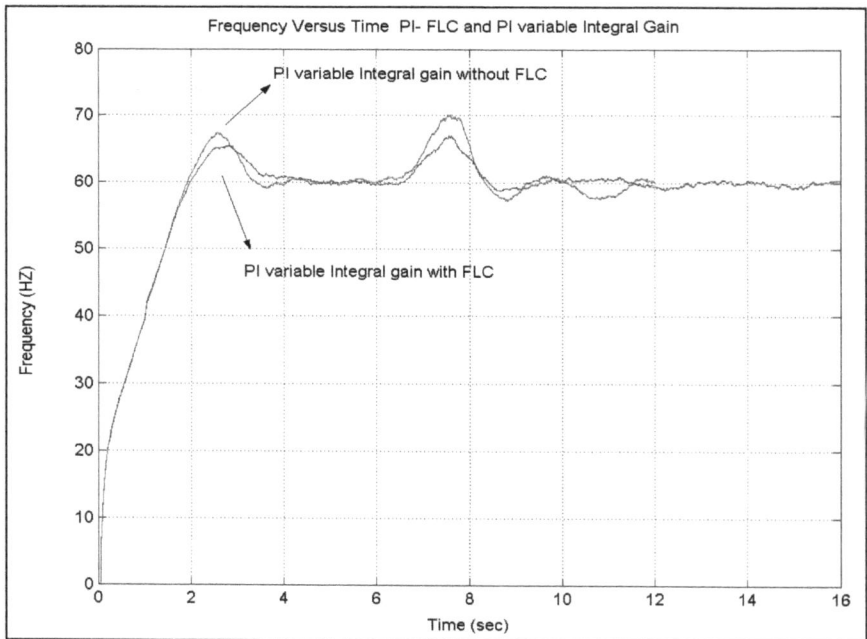

Figure 15.10: Dynamic Response of Stator Frequency for PI with and Without FLC.

ANN_FLC Simulation Results

The simulation program is carried out for variable learning rates(α_{rv}, α_{rf}). It is noticed from the simulation results that the value of percentage overshoot, rising time and settling time change as α_{rv} is changed as seen in Figures 15.11(a-c) and Figure 15.12 show the membership function of error, change of error and output respectively. Table 15.2 shows the lookup table of fuzzy set rules for reactive power control. Figure 15.13 shows the comparison between terminal voltage versus time for ANN performance with and without fuzzy set technique to compute the variable learning rate. Figure 15.14 shows the comparison between stator frequency versus time for ANN performance with and without fuzzy set technique.

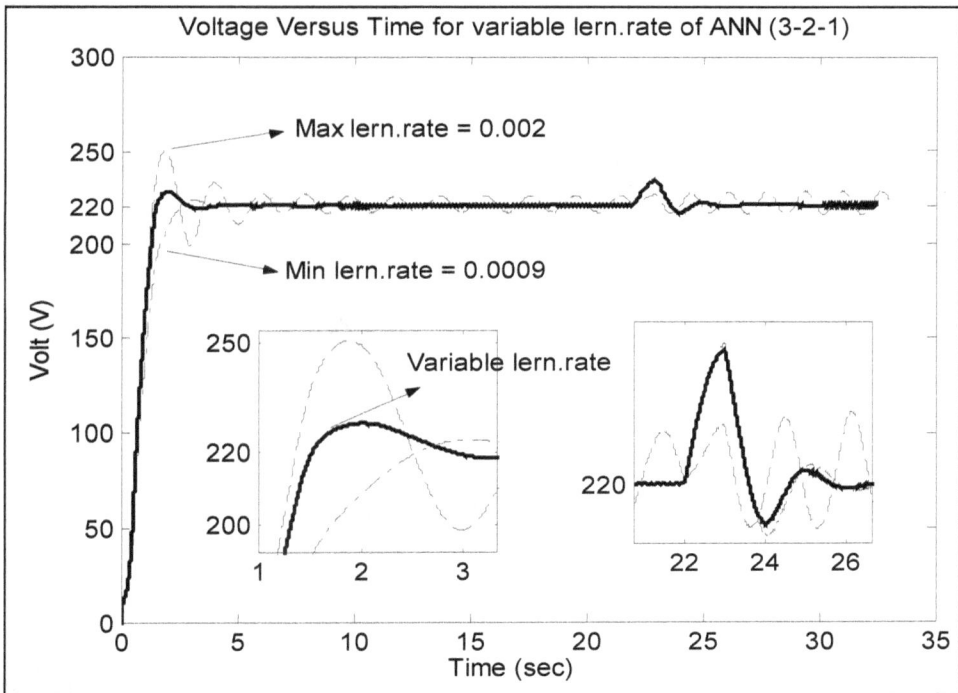

Figure 15.11a: Dynamic Response of the Terminal Voltage with α_{rv}.

Table 15.2: Fuzzy Set Rules for Reactive Power Control by ANN

Ve'_change of Error					
HB_V	HS_V	AV_V	LS_V	LB_V	
HS_V	AV_V	LS_V	LB_V	LB_V	LB_V
HS_V	AV_V	LS_V	LS_V	LB_V	LS_V
HS_V	HS_V	AV_V	LS_V	LS_V	AV_V
HB_V	HB_V	HS_V	AV_V	AV_V	HS_V
HB_V	HB_V	HS_V	HS_V	HS_V	HB_V

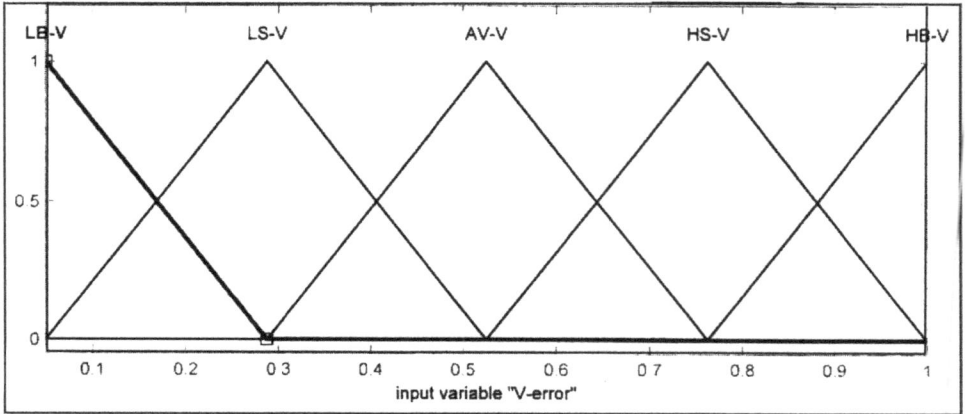

Figure 15.11b: Membership Function of Voltage Error.

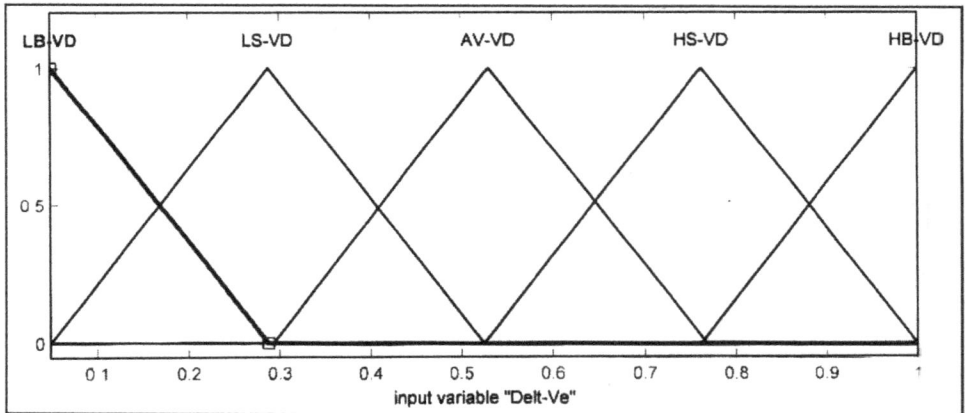

Figure 15.11c: Membership Function of Change in Voltage Error.

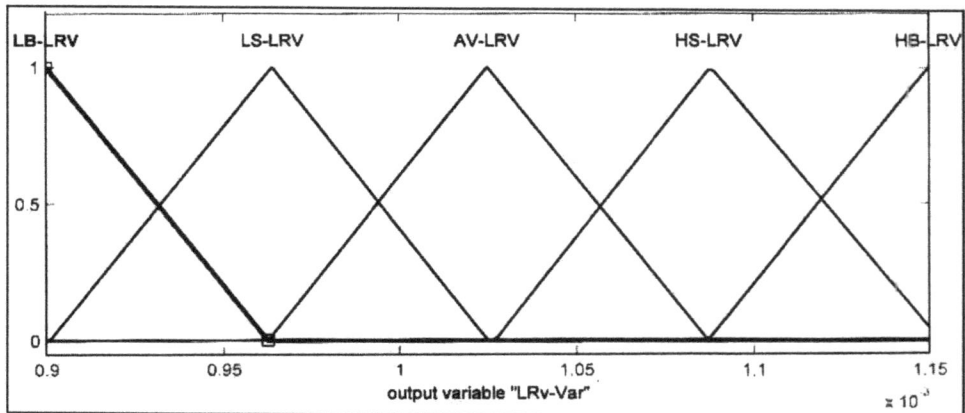

Figure 15.12: Membership Function of Learning Rate.

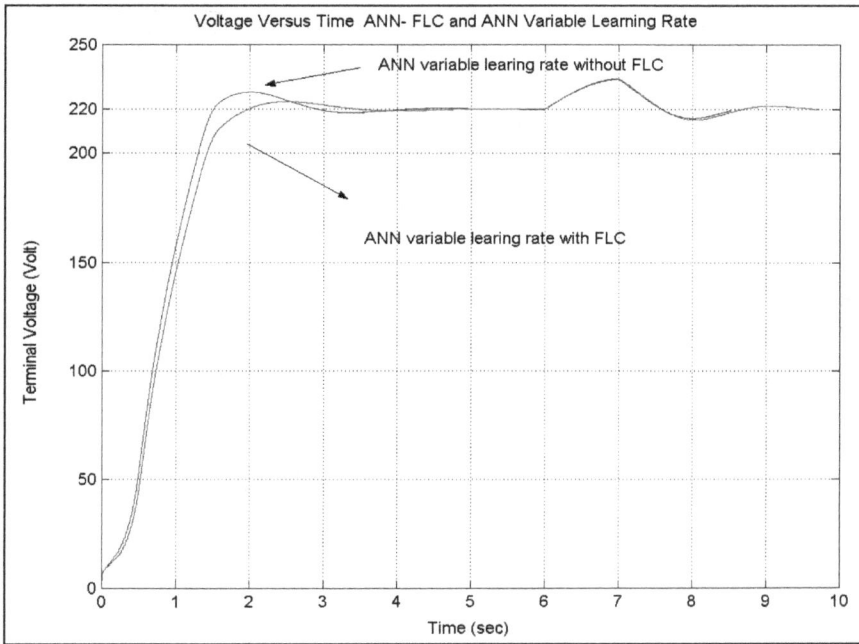

Figure 15.13: Dynamic Response of Terminal Voltage for ANN with and Without FLC.

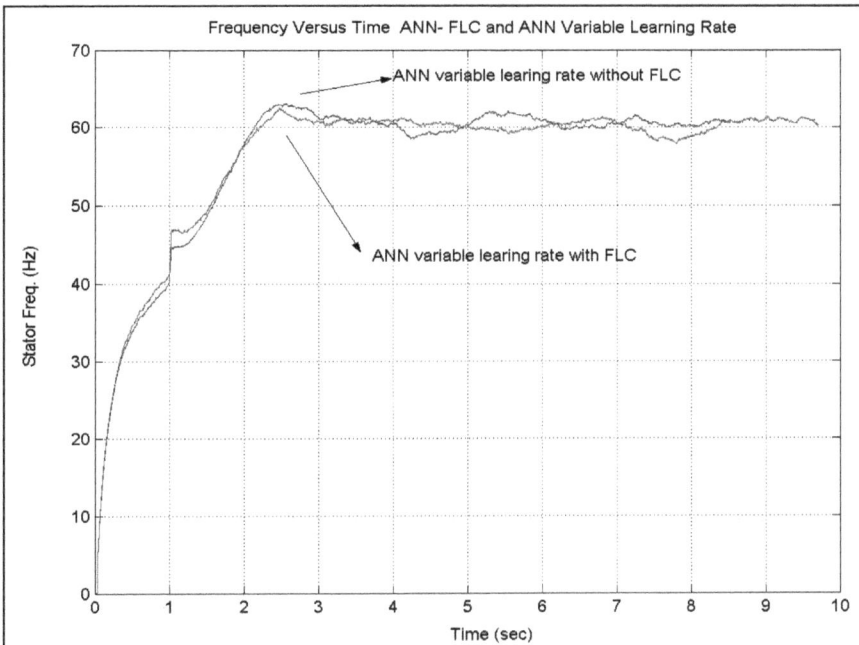

Figure 15.14: Dynamic Response of Stator Frequency for ANN with and Without FLC.

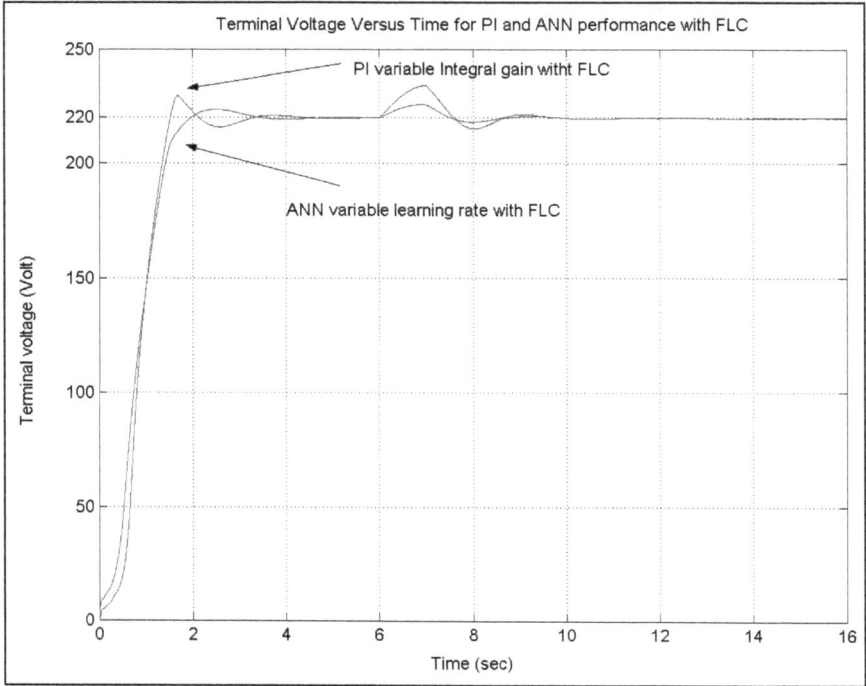

Figure 15.15: Dynamic Response of Terminal Voltage for PI and ANN with FLC.

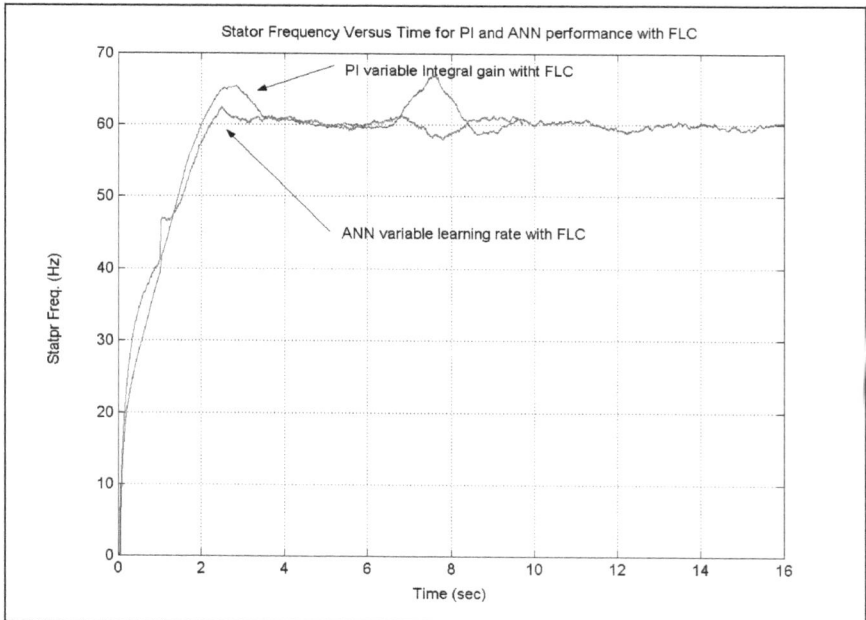

Figure 15.16: Dynamic Response of Stator Frequency for PI and ANN with FLC.

Finally a comparison between PI_FLC and ANN_FLC is madein Figures 15.15 and 15.16 for terminal voltage and stator frequency, respectively. The best performance is found for ANN-FLC technique.

Conclusions

This paper discussed the enhancement of the performance for the system of SEIG driven by WTusing various controllers as PI_FLC, ANN_FLC controllers. The best performance is found with ANN_FLC technique.

References

1   Li, Wang, and Jian-Yi-SU, "Dynamic Performance of An Isolated Self Excited Induction Generator Under Various Loading Conditions", *IEEE Transactions on Energy Conversion*, Vol. 15, No.1, March 1999, pp. 93-100.

2   Li, Wang, and Ching- Huei Lee, "Long- Shunt and Short- Shunt Connections on Dynamic Performance of a SEIG Feeding an Induction Motor Load", *IEEE Transactions on Energy Conversion*, Vol. 14, No.1, 2000, pp. 1-7.

3   Mokhymar, Sabry. M. Aly, "Genetic Algorithms Based Control System Design of Self-Excited Induction Generator", *Acta Polytechnica*, Vo.46 No. 2/2006, pp. 11-22.

4   Ezzeldin S. Abdin and Wilson Xu, "Control Design and Dynamic Performance Analysis of a Wind Turbine- Induction Generator Unit", *IEEE Transaction on Energy Conversion*, Vol. 15, No.1, March 2000, pp. 91-96.

5   Ahmed M. Atallah and Adel Ahmed "Terminal Voltage Control of Slef Excited Induction Generators" Sixth Middle-East Power Systems Conference (MEPCON'98), Mansoura, Egypt, Dec. 15-17,1998, pp. 110-118.

6   C. Marduchus, "Switched Capacitor Circuits for reactive power generation ",Ph.D. Thesis, Brunuel University, 1983.

7   Hussein F. Soliman, W. Helmy and M.A.L.Badr " ANN-Based Controller For Excitation Control Of A Two-Machine Power System.", *Sci. Bull. Fac.Eng. AinShamsUniversity*. Part II. Vol.38,No.3, Sep. 30, 03, pp. 503-517.

8   Mokhymar, Sabry. M. Aly., "Dynamic Performance Enhancement of Self Excited Induction Generator Driven By Wind Energy Using ANN Controllers ", Scientific Bulletin in Ain shams UN. Faculty of Engineering June, 2004 titled.

9   Mokhymar, Sabry. M. Aly, "dynamic performance Improvement of Wind-driven Induction generator using fuzzy–neural controller", 2nd international conference for scientific research and its application, Cairo, December 2005.

10   Xianzhong Cui and Kang G., "Application of Neural Networks to Temperature Control in Thermal Power Plants", Engng. Appl. Vol. 5. No. 6. pp. 527-538, 1992, Printed in Great Britain- Per. Press Ltd, pp. 527- 538.

11   Mokhymar, Sabry. M. Aly, "Fuzzy Algorithm for supervisory control of Self excited induction generator. ", *Acta Polytechnica*, Accepted for publication year of 2006.

12   Kevin M. Passino, Stephen Yurkovich "Fuzzy Control" Department of Electrical Engineering, the Ohio State University, Library of Congress Cataloging-in-Publication Data, Passino, Kevin M. Fuzzy control/Kevin M. Passino and Stephen Yurkovich., p. cm. Includes bibliographical references and index. ISBN 0-201-18074-X

13   Mokhymar, Sabry. M. Aly "Dynamic Performance Improving of Wind – Driven Induction Generator Using ANN and PI", 2ndinternational conference for Environmental Engineering, Cairo, April 2007.

Appendix - A

v_{ds}: Stator Voltage 's (volt) Differential Equation at Direct Axis (A1)

$$V_{ds} = -R_s \cdot i_{ds} - \left(\frac{\omega}{\omega_b}\right)\varphi_{qs} + p\left(\frac{\varphi_{ds}}{\omega_b}\right)$$

v_{qs}: Stator Voltage 's Differential Equation at Quadrate Axis (A2)

$$V_{qs} = -R_s \cdot i_{qs} + \left(\frac{\omega}{\omega_b}\right)\varphi_{ds} + p\left(\frac{\varphi_{qs}}{\omega_b}\right)$$

v_{dr}: Rotor Voltage 's Differential Equation at Direct Axis (A3)

$$V_{dr} = R_r \cdot i_{dr} - \left(\frac{(\omega - \omega_r)}{\omega_b}\right)\varphi_{qr} + p\left(\frac{\varphi_{dr}}{\omega_b}\right)$$

v_{qr}: Rotor Voltage 's Differential Equation at Quadrate Axis: (A4)

$$V_{qr} = R_r \cdot i_{qr} + \left(\frac{(\omega - \omega_r)}{\omega_b}\right)\varphi_{dr} + p\left(\frac{\varphi_{qr}}{\omega_b}\right)$$

$$C_p = \left[\left(0.44 - 0.0167\ \beta\right)Sin\left(\frac{\pi(\mu - 3)}{(15 - 0.3\beta)}\right) - 0.00184(\mu - 3)\beta\right]$$

(A5)

where,

ω_m is the mechanical speed (rad/s); P_m is the mechanical power (kW), T_m is the mechanical torque (nm), n is the rotor revolution per minute (rpm), C_p is the power coefficient of the wind turbine, β is the blade pitch angle (degree), μ is the tip speed ratio, V_w is the wind speed (m/s), D is the the rotor Diameter (m) of the wind turbine, $\pi = 3.14$ and $\rho = $ Air density (kg/m³)

Output of Output Layer Neuron

$$N = (F_c \times W_{Aold}) + (F_d \times W_{Bold}) + B$$

(A6)

where,

B is the bias of the output layer, Fc is output of Input Layer Logistic Function for second Neuron c, Fd is output of Input Layer Logistic Function for second Neuron d, W_{aold} is initial condition of First Weight (W_A) Calculation of Hidden Layer, W_{bold} is initial condition of second Weight (W_B) Calculation of Hidden Layer

Output of Output Layer Logistic Function for output of ANN:

$$Z = \left[\frac{1}{1 + \exp(K \times N)} \right]$$

(A7)

Ve_ error	Ve'_Change of Error		
	LO	AV	HI
LO	LO	LO	LO
AV	AV	AV	AV
HI	HI	HI	HI

Ve'_Change of Error						Ve_ error
HB_V	HS_V	AV_V	LS_V	LB_V		
HS_V	AV_V	LS_V	LB_V	LB_V	LB_V	
HS_V	AV_V	LS_V	LS_V	LB_V	LS_V	
HS_V	HS_V	AV_V	LS_V	LS_V	AV_V	
HB_V	HB_V	HS_V	AV_V	AV_V	HS_V	
HB_V	HB_V	HS_V	HS_V	HS_V	HB_V	

Ahmedabad Declaration on

Energy Models in Emerging Economies – Post COP 21

WE, THE DELEGATES to the 3 days international workshop on "Evolving Energy Models in Emerging Economies – post COP 21", jointly organised by the Centre for Science and Technology of the Non-Aligned and Other Developing Countries (NAM S&T Centre) and the Society of Energy Engineers and Managers (SEEM), India, and supported by the Gujarat Technical University and International Copper Association India as knowledge partners, at Ahmedabad, India, during 12th– 14th December 2016,

COMPRISING Scientists, Academicians, Technocrats, Engineers, Consultants, Industrialists, Policy Makers, Energy Managers and Energy Auditors from Cuba, Egypt, India, Indonesia, Iran, Malaysia, Nigeria, Sri Lanka, Togo, Turkey, the USA, Vietnam, Zambia and Zimbabwe;

BELIEVING THAT the Intended Nationally Determined Contributions declared by the global communities at COP 21 included quantified emission reduction targets as well as renewable energy targets; and that meaningful and effective achievement of these targets demands proactive, intensified and harmonious efforts; and

APPRECIATING THAT diversification to new, renewable and sustainable energy sources have opened up opportunities with associated risks, that demand significant market transformation to be ably backed up by government mechanisms that will ensure energy access, affordability and security to the people of the nations;

UNANIMOUSLY RESOLVE THAT:

★ New business models of energy aided by renewable energy technologies and energy efficient practices need to be identified and promoted in all countries, which should ensure continual reduction of carbon foot print of the nations and enhance the energy infrastructure. Such models should have lower dependence on the government subsidies and are sustainable. This may be similar to the Blockchain Technology, an initiative where prosumers (consumers who also produce energy) of distributed generation share their excess/deficit supply/demand in a cooperative model.

★ Weak infrastructure resilience being identified as a major challenge to transition to new energy models, governments may take urgent policy decisions and regulatory measures to ease the transition and handhold the new energy entrepreneurs. Non-governmental organisations in the energy and environment sectors too may play a catalytic role in this mission.

★ Appropriate waste management practices should be followed for solar photovoltaic, other renewable energy systems and energy efficient devices at the end of their useful life. The material recovery from waste and waste-to-energy conversion should be promoted in all countries.

★ Eco-friendly energy technologies having zero impact on air, land and water resources should be implemented and the experience be widely shared.

★ Energy audit should be made mandatory in all countries, and there should be suitable mechanisms to ensure implementation of audit recommendations.

★ Green building construction should be promoted by governmental incentives in all countries.

★ Since developing and managing Evolving Energy Models require necessary capacity building and resource allocation, the planners should come up with feasible economic measures to support the same. The developed nations should also adhere to the commitments of finance in the Paris Agreement COP - 21.

★ The informal, traditional, micro, small and medium scale manufacturing sectors should be transformed to become environmentally and financially sustainable by fostering R&D, capacity building and skill upgradation so that they are globally competitive.

★ Since a large section of masses in the emerging economies does not have adequate supply of commercial energy, appropriate planning and implementation of the new energy models of distributive nature should be adopted. Large scale involvement of middle level entrepreneurs should be ensured with the evolving energy models like Energy Service Companies (ESCO) and Renewable Energy Supply Companies (RESCO). People's participation, including those of NGOs, should be encouraged at the grassroot level by appropriate government policies.

★ Apart from energy efficiency there should be an integrated approach focusing on resource efficiency, cleaner production practices and different forms of energy.

★ Climate change departments should be set up in the governments of all countries at both the federal/national and state levels, emulating the initiative of the government of Gujarat in India. Appropriate tools and services to monitor climate change and to assess the effectiveness of climate change abatement initiatives should be developed in all countries.

★ Emphasis should be given to continual cooperation among developing countries to share best practices in matters related to energy efficiency and climate change.

The Society of Energy Engineers and Managers (SEEM), a non-profit organisation of certified energy auditors and managers in India, urged other countries to set up similar organisations to carry forward the resolutions of this workshop. The delegates of the Ahmedabad workshop from the following countries expressed willingness to setup similar organisations in their countries with the support of SEEM – India:

★ Egypt
★ Iran
★ Nigeria
★ Sri Lanka
★ Togo
★ Turkey

This declaration calls upon the governments, business communities and energy stakeholders to act on their obligation to provide a sustainable living, meeting the developmental aspirations of the people of the nations, that also ensures a clean, green and safe planet to the present and future generations.

THUS RESOLVED AND ADOPTED ON THE 14THDECEMBER 2016 AT AHMEDABAD, GUJARAT, INDIA